Indian Hut on the Mazaruni River

WILLIAM BEEBE
AUTHOR OF "EDGE OF THE JUNGLE," ETC.

JUNGLE DAYS

WITH SEVEN ILLUSTRATIONS

GARDEN CITY, NEW YORK
GARDEN CITY PUBLISHING CO., INC.

Copyright, 1923
by
The Atlantic Monthly Co., Inc.

Copyright, 1925
by
The Curtis Publishing Co.

Copyright, 1925
by
William Beebe

Made in the United States of America

CONTENTS

CHAPTER	PAGE
I.—A Chain of Jungle Life	3
II.—My Jungle Table	26
III.—A Midnight Beach Combing	49
IV.—Falling Leaves	71
V.—The Jungle Sluggard	92
VI.—Mangrove Mystery	113
VII.—The Life of Death	137
VIII.—Old-Time People	166
IX.—The Bird of the Wine-Colored Egg	182
Index	203

JUNGLE DAYS

Jungle Days

I

A CHAIN OF JUNGLE LIFE

*This is the story of Opalina
Who lived in the Tad,
Who became the Frog,
Who was eaten by Fish,
Who nourished the Snake,
Who was caught by the Owl,
But fed the Vulture,
Who was shot by Me,
Who wrote this Tale,
Which the Editor took,
And published it Here,
To be read by You,
The last in The Chain,
Of Life in the tropical Jungle.*

I OFFER a living chain of ten links—the first a tiny delicate being, one hundred to the inch, deep in the jungle, with the strangest home in the world—my last, you the present reader of these lines. Between, there befell certain things, of which I attempt falteringly to write. To know and think them is very worth while, to have discovered them is sheer joy, but to write of them is impertinence, so exciting and unreal are they in reality, and so tame and humdrum are any combinations of our twenty-six letters.

Somewhere today a worm has given up existence, a mouse has been slain, a spider snatched from the web, a jungle bird torn sleeping from its perch; else we should have no song of robin, nor flash of reynard's red, no humming flight of wasp, nor grace of crouching ocelot. In tropical jungles, in Northern home orchards, anywhere you will, unnumbered activities of bird and beast and insect require daily toll of life.

Now and then we actually witness one of these tragedies or successes—whichever point of view we take—appearing to us as an exciting but isolated event. When once we grasp the idea of chains of life, each of these occurrences assumes a new meaning. Like everything else in the world it is not isolated, but closely linked with other similar happenings. I have sometimes traced even closed chains, one of the shortest of which consisted of predacious flycatchers which fed upon young lizards of a species which, when it grew up, climbed trees and devoured the nestling flycatchers!

One of the most wonderful zoological "Houses that Jack built," was this of Opalina's, a long,

A CHAIN OF JUNGLE LIFE

swinging, exciting chain, including in its links a Protozoan, two stages of Amphibians, a Fish, a Reptile, two Birds and (unless some intervening act of legislature bars the fact as immoral and illegal) three Mammals,—myself, the Editor, and You.

As I do not want to make it into a mere imaginary animal story, however probable, I will begin, like Dickens, in the middle. I can cope, however lamely, with the entrance and participation of the earlier links, but am wholly out of my depth from the time when I mail my tale. The Akawai Indian who took it upon its first lap toward the Editor should by rights have a place in the chain, especially when I think how much better he might tell of the interrelationships of the various links than can I. Still, I know the shape of the owl's wings when it dropped upon the snake, but I do not know why the Editor accepted this; I can imitate the death scream of the frog when the fish seized it, but I have no idea why You purchased this volume nor whether you perceive in my tale the huge bed of ignorance in which I have planted this scanty crop of facts. Nor do I know the future of this book, whether it will go to the garret, to be ferreted out in future years by other links, as I used to do, or whether it will find its way to mid-Asia or the

JUNGLE DAYS

Malay States, or, as I once saw a magazine, half buried, like the pyramids, in Saharan sands, where it had slipped from the camel load of some unknown traveller.

I left my Kartabo laboratory one morning with my gun, headed for the old Dutch stelling. Happening to glance up I saw a mote, lit with the oblique rays of the morning sun. The mote drifted about in circles, which became spirals; the mote became a dot, then a spot, then an oblong, and down the heavens from unknown heights, with the whole of British Guiana spread out beneath him from which to choose, swept a vulture into my very path. We had a quintet, a small flock of our own vultures who came sifting down the sky, day after day, to the feasts of monkey bodies and wild peccaries which we spread for them. I knew all these by sight, from one peculiarity or another, for I was accustomed to watch them hour after hour, striving to learn something of that wonderful soaring, of which all my many hours of flying had taught me nothing.

This bird was a stranger, perhaps from the coast or the inland savannas, for to these birds great spaces are only matters of brief moments. I wanted a yellow-headed vulture, both for the painting of its marvellous head colors, and for the

A CHAIN OF JUNGLE LIFE

strange, intensely interesting, one-sided, down-at-the-heel syrinx, which, with the voice, had dissolved long ages ago, leaving only a whistling breath, and an irregular complex of bones straggling over the windpipe. Some day I shall dilate upon vultures as pets—being surpassed in cleanliness, affectionateness and tameness only by baby bears, sloths and certain monkeys.

But today I wanted the newcomer as a specimen. I was surprised to see that he did not head for the regular vulture table, but slid along a slant of the east wind, banked around its side, spreading and curling upward his wing-finger-tips and finally resting against its front edge. Down this he sank slowly, balancing with the grace of perfect mastery, and again swung round and settled suddenly down shore, beyond a web of mangrove roots. This took me by surprise, and I changed my route and pushed through the undergrowth of young palms. Before I came within sight, the bird heard me, rose with a whipping of great pinions and swept around three-fourths of a circle before I could catch enough of a glimpse to drop him. The impetus carried him on and completed the circle, and when I came out on the Cuyuni shore I saw him spread out on what must have been the exact spot from which he had risen.

JUNGLE DAYS

I walked along a greenheart log with little crabs scuttling off on each side, and as I looked ahead at the vulture I saw to my great surprise that it had more colors than any yellow-headed vulture should have, and its plumage was somehow very different. This excited me so that I promptly slipped off the log and joined the crabs in the mud. Paying more attention to my steps I did not again look up until I had reached the tuft of low reeds on which the bird lay. Now at last I understood why my bird had metamorphosed in death, and also why it had chosen to descend to this spot. Instead of one bird, there were two, and a reptile. Another tragedy had taken place a few hours earlier, before dawn, a double death, and the sight of these three creatures brought to mind at once the chain for which I am always on the lookout. I picked up my chain by the middle and began searching both ways for the missing links.

The vulture lay with magnificent wings outspread, partly covering a big, spectacled owl, whose dishevelled plumage was in turn wrapped about by several coils of a moderate-sized anaconda. Here was an excellent beginning for my chain, and at once I visualized myself and the snake, although alternate links, yet coupled in contradistinction to my editor and the vulture, the first two

A CHAIN OF JUNGLE LIFE

having entered the chain by means of death, whereas the vulture had simply joined in the pacifistic manner of its kind, and as my editor has dealt gently with me heretofore, I allowed myself to believe that his entrance might also be through no more rough handling than a blue slip.

The head of the vulture was already losing some of its brilliant chrome and saffron, so I took it up, noted the conditions of the surrounding sandy mud, and gathered together my spoils. I would have passed within a few feet of the owl and the snake and never discovered them, so close were they in color to the dark reddish beach, yet the vulture with its small eyes and minute nerves had detected this tragedy when still perhaps a mile high in the air, or half a mile up-river. There could have been no odor, nor has the bird any adequate nostrils to detect it, had there been one. It was sheer keenness of vision. I looked at the bird's claws and their weakness showed the necessity of the eternal search for carrion or recently killed creatures. Here in a half minute, it had devoured an eye of the owl and both of those of the serpent. It is a curious thing, this predilection for eyes; give a monkey a fish, and the eyes are the first titbits taken.

Through the vulture I come to the owl link, a

splendid bird clad in the colors of its time of hunting; a great, soft, dark, shadow of a bird, with tiny body and long fluffy plumage of twilight buff and ebony night, lit by twin, orange moons of eyes. The name "spectacled owl" is really more applicable to the downy nestling which is like a white powder puff with two dark feathery spectacles around the eyes. Its name is one of those which I am fond of repeating rapidly—*Pulsatrix perspicillata perspicillata*. Etymologies do not grow in the jungle and my memory is noted only for its consistent vagueness, but if the owl's title does not mean *The Eye-browed One Who Strikes*, it ought to, especially as the subspecific trinomial grants it two eyebrows.

I would give much to know just what the beginning of the combat was like. The middle I could reconstruct without question, and the end was only too apparent. By a most singular coincidence, a few years before, and less than three miles away, I had found the desiccated remains of another spectacled owl mingled with the bones of a snake, only in that instance, the fangs indicated a small fer-de-lance, the owl having succumbed to its venom. This time the owl had rashly attacked a serpent far too heavy for it to lift, or even, as it turned out, successfully to battle with. The mud

A CHAIN OF JUNGLE LIFE

had been churned up for a foot in all directions, and the bird's plumage showed that it must have rolled over and over. The anaconda, having just fed, had come out of the water and was probably stretched out on the sand and mud, as I have seen them, both by full sun and in the moonlight. These owls are birds rather of the creeks and river banks than of the deep jungle, and in their food I have found shrimps, crabs, fish and young birds. Once a few snake vertebræ showed that these reptiles are occasionally killed and devoured.

Whatever possessed the bird to strike its talons deep into the neck and back of this anaconda, none but the owl could say, but from then on the story was written by the combatants and their environment. The snake, like a flash, threw two coils around bird, wings and all, and clamped these tight with a cross-vise of muscle. The tighter the coils compressed the deeper the talons of the bird were driven in, but the damage was done with the first strike, and if owl and snake had parted at this moment, neither could have survived. It was a swift, terrible and short fight. The snake could not use its teeth and the bird had no time to bring its beak into play, and there in the night, with the lapping waves of the falling tide only two or three feet away, the two creatures of prey met and fought

JUNGLE DAYS

and died, in darkness and silence, locked fast together.

A few nights before I had heard, on the opposite side of the bungalow, the deep, sonorous cry of the spectacled owl; within the week I had passed the line-and-crescents track of anacondas, one about the size of this snake and another much larger. And now fate had linked their lives, or rather deaths, with my life, using as her divining rod, the focussing of a sky-soaring vulture.

The owl had not fed that evening, although the bird was so well nourished that it could never have been driven to its foolhardy feat by stress of hunger. Hopeful of lengthening the chain, I rejoiced to see a suspicious swelling about the middle of the snake, which dissection resolved into a good-sized fish—itself carnivorous, locally called a basha. This was the first time I had known one of these fish to fall a victim to a land creature, except in the case of a big kingfisher who had caught two small ones. Like the owl and anaconda, bashas are nocturnal in their activities, and, according to their size, feed on small shrimps, big shrimps, and so on up to six or eight inch catfish. They are built on swift, torpedo-like lines, and clad in iridescent silver mail.

From what I have seen of the habits of ana-

A CHAIN OF JUNGLE LIFE

condas, I should say that this one had left its hole high up among the upper beach roots late in the night, and softly wound its way down into the rising tide. Here after drinking, the snake sometimes pursues and catches small fish and frogs, but the usual method is to coil up beside a half-buried stick or log and await the tide and the manna it brings. In the van of the waters comes a host of small fry, followed by their pursuers or by larger vegetable feeders, and the serpent has but to choose. In this mangrove lagoon then, there must have been a swirl and a splash, a passive holding fast by the snake for a while until the right opportunity offered, and then a swift throw of coils. There must then be no mistake as to orientation of the fish. It would be a fatal error to attempt the tail first, with scales on end and serried spines to pierce the thickest tissues. It is beyond my knowledge how one of these fish can be swallowed even head first without serious laceration. But here was optical proof of its possibility, a newly swallowed basha, so recently caught that he appeared as in life, with even the delicate turquoise pigment beneath his scales, acting on his silvery armor as quicksilver under glass. The tooth marks of the snake were still clearly visible on the scales,—another link, going steadily down the classes of vertebrates, mammal, bird, reptile and fish, and still my magic boxes were unexhausted.

Excitedly I cut open the fish. An organism more unlike that of the snake would be hard to imagine. There I had followed an elongated stomach, and had left unexplored many feet of alimentary canal. Here, the fish had his heart literally in his mouth, while his liver and lights were only a very short distance behind, followed by a great expanse of tail to wag him at its will, and drive him through the water with the speed of twin propellers. His eyes are wonderful for night hunting, large, wide, and bent in the middle so he can see both above and on each side. But all this wide-angled vision availed nothing against the lidless, motionless watch of the ambushed anaconda. Searching the crevices of the rocks and logs for timorous small fry, the basha had sculled too close, and the jaws which closed upon him were backed by too much muscle, and too perfect a throttling machine to allow of the least chance of escape. It was a big basha compared with the moderate-sized snake but the fierce eyes had judged well, as the evidence before me proved.

Still my chain held true, and in the stomach of the basha I found what I wanted—another link, and more than I could have hoped for—a repre-

A CHAIN OF JUNGLE LIFE

sentative of the fifth and last class of vertebrate animals living on the earth, an Amphibian, an enormous frog. This too had been a swift-forged link, so recent that digestion had only affected the head of the creature. I drew it out, set it upon its great squat legs, and there was a grandmother frog almost as in life, a Pok-poke as the Indians call it, or, as a herpetologist would prefer, *Leptodactylus caliginosus*,—the Smoky Jungle Frog.

She lived in the jungle just behind, where she and a sister of hers had their curious nests of foam, which they guarded from danger, while the tadpoles grew and squirmed within its sudsy mesh as if there were no water in the world. I had watched one of the two, perhaps this one, for hours, and I saw her dart angrily after little fish which came too near. Then, this night, the high full-moon tides had swept over the barrier back of the mangrove roots and set the tadpoles free, and the mother frogs were at liberty to go where they pleased.

From my cot in the bungalow to the south, I had heard in the early part of the night, the death scream of a frog, and it must have been at that moment that somehow the basha had caught the great amphibian. This frog is one of the fiercest of its class, and captures mice, reptiles and small fish without trouble. It is even cannibalistic on very slight provocation, and two of equal size will sometimes endeavor to swallow one another in the most appallingly matter-of-fact manner.

They represent the opposite extreme in temperament from the pleasantly philosophical giant toads. In outward appearance in the dim light of dusk, the two groups are not unlike, but the moment they are taken in the hand all doubt ceases. After one dive for freedom the toad resigns himself to fate, only venting his spleen in much puffing out of his sides, while the frog either fights until exhausted, or pretends death until opportunity offers for a last mad dash.

In this case the frog must have leaped into the deep water beyond the usual barrier and while swimming been attacked by the equally voracious fish. In addition to the regular croak of this species, it has a most unexpected and unamphibian yell or scream, given only when it thinks itself at the last extremity. It is most unnerving when the frog, held firmly by the hind legs, suddenly puts its whole soul into an ear-splitting *peent! peent! peent! peent! peent!*

Many a time they are probably saved from death by this cry which startles like a sudden blow, but tonight no utterance in the world could have saved it; its assailant was dumb and all but deaf to

A CHAIN OF JUNGLE LIFE

aerial sounds. Its cries were smothered in the water as the fish dived and nuzzled it about the roots, as bashas do with their food,—and it became another link in the chain.

Like a miser with one unfilled coffer, or a gambler with an unfilled royal flush, I went eagerly at the frog with forceps and scalpel. But beyond a meagre residuum of eggs, there was nothing but shrunken organs in its body. The rashness of its venture into river water was perhaps prompted by hunger after its long maternal fast while it watched over its egg-filled nest of foam.

Hopeful to the last, I scrape some mucus from its food canal, place it in a drop of water under my microscope, and—discover Opalina, my last link, which in the course of its most astonishing life history gives me still another.

To the naked eye there is nothing visible—the water seems clear, but when I enlarge the diameter of magnification I lift the veil on another world, and there swim into view a dozen minute lives, oval little beings covered with curving lines, giving the appearance of wandering finger prints. In some lights these are iridescent and they then will deserve the name of Opalina. As for their personality, they are oval and rather flat, it would take one hundred of them to stretch an inch, they have no mouth, and they are covered with a fur of flagella with which they whip themselves through the water. Indeed the whole of their little selves consists of a multitude of nuclei, sometimes as many as two hundred, exactly alike,—facial expression, profile, torso, limbs, pose, all are summed up in rounded nuclei, partly obscured by a mist of vibrating flagella.

As for their gait, they move along with colorful waves, steadily and gently, not keeping an absolutely straight course and making rather much leeway, as any rounded, keelless craft, surrounded with its own paddle-wheels, must expect to do.

I have placed Opalina under very strange and unpleasant conditions in thus subjecting it to the inhospitable qualities of a drop of clear water. Even as I watch, it begins to slow down, and the flagella move less rapidly and evenly. It prefers an environment far different, where I discovered it living happily and contentedly in the stomach and intestines of a frog, where its iridescence was lost, or rather had never existed in the absolute darkness; where its delicate hairs must often be unmercifully crushed and bent in the ever-moving tube, and where air and sky, trees and sun, sound and color were forever unknown; in their place only bits of half-digested ants and beetles, thousand-legs and worms, rolled and tumbled along in the dense gastric stream of acid pepsin; a strange choice of home for one of our fellow living beings on the earth.

After an Opalina has flagellated itself about, and fed for a time in its strange, almost crystalline way on the juices of its host's food, its body begins to contract, and narrows across the center until it looks somewhat like a map of the New World. Finally its isthmus thread breaks and two Opalinas swim placidly off, both identical, except that they have half the number of nuclei as before. We cannot wonder that there is no backward glance, or wave of cilia, or even memory of their other body, for they are themselves, or rather it is **they**, or it is each: our whole vocabulary, our entire stock of pronouns breaks down, our very conception of individuality is shattered by the life of Opalina.

Each daughter cell or self-twin, or whatever we choose to conceive it, divides in turn. Finally there comes a day (or rather some Einstein period of space-time, for there are no days in a frog's stomach!) when Opalina's fraction has reached a stage with only two nuclei. When this has creased and stretched, and finally broken like two bits of drawn-out molasses candy, we have the last divisional possibility. The time for the great adven-

JUNGLE DAYS

ture has arrived, with decks cleared for action, or, as a protozoölogist would put it, with the flagellate's protoplasm uni-nucleate, approximating encystment.

The encysting process is but slightly understood, but the tiny one-two-hundredth-of-its-former-self Opalina curls up, its paddle-wheels run down, it forms a shell, and rolls into the current which it has withstood for a Protozoan's lifetime. Out into the world drifts the minute ball of latent life, a plaything of the cosmos, permitted neither to see, hear, eat, nor to move of its own volition. It hopes (only it cannot even desire) to find itself in water, it must fall or be washed into a pool with tadpoles, one of which must come along at the right moment and swallow it with the débris upon which it rests. The possibility of this elaborate concatenation of events has everything against it, and yet it must occur or death will result. No wonder that the population of Opalinas does not overstock its limited and retired environment!

Supposing that all happens as it should, and that the only chance in a hundred thousand comes to pass, the encysted being knows or is affected in some mysterious way by entrance into the body of the tadpole. The cyst is dissolved and the infant Opalina begins to feed and to develop new nuclei.

A CHAIN OF JUNGLE LIFE

Like the queen ant after she has been walled forever into her chamber, the life of the little Onecell would seem to be extremely sedentary and humdrum, in fact monotonous, until its turn comes to fractionize itself, and again severally to go into the outside world, multiplied and by installments. But as the queen ant had her one superlative day of sunlight, heavenly flight and a mate, so Opalina, while she is still wholly herself, has a little adventure all her own.

Let us strive to visualize her environment as it would appear to her if she could find time and ability, with her single cell, to do more than feed and bisect herself. Once free from her horny cyst she stretches her drop of a body, sets all her paddle-hairs in motion and swims slowly off. If we suppose that she has been swallowed by a tadpole an inch long, her living quarters are astonishingly spacious or rather elongated. Passing from end to end she would find a living tube two feet in length, a dizzy path to traverse, as it is curled in a tight, many-whorled spiral,—the stairway, the domicile, the universe at present for Opalina. She is compelled to be a vegetarian, for nothing but masses of decayed leaf tissue and black mud and algæ come down the stairway. For many days there is only the sound of water gurgling past the

JUNGLE DAYS

tadpole's gills, or glimpses of sticks and leaves and the occasional flash of a small fish through the thin skin periscope of its body.

Then the tadpole's mumbling even of half-rotted leaves comes to an end, and both it and its guests begin to fast. Down the whorls comes less and less of vegetable detritus, and Opalina must feel like the crew of a submarine when the food supply runs short. At the same time something very strange happens, the experience of which eludes our utmost imagination. Poe wrote a memorable tale of a prison cell which day by day grew smaller, and Opalina goes through much the same adventure. If she frequently traverses her tube, she finds it growing shorter and shorter. As it contracts, the spiral untwists and straightens out, while all the time the rations are cut off. A dark curtain of pigment is drawn across the epidermal periscope and as books of dire adventure say, the 'horror of darkness is added to the terrible mental uncertainty.' The whole movement of the organism changes; there is no longer the rush and swish of water, and the even, undulatory motion alters to a series of spasmodic jerks,—quite the opposite of ordinary transition from water to land. Instead of water rushing through the gills of her host, Opalina might now hear strange musical sounds,

A CHAIN OF JUNGLE LIFE

loud and low, the singing of insects, the soughing of swamp palms.

Opalina about this time, should be feeling very low in her mind from lack of food, and the uncertainty of explanation of why the larger her host grew, the smaller, more confined became her quarters. The tension is relieved at last by a new influx of provender, but no more inert mold or disintegrated leaves. Down the short, straight tube appears a live millipede, kicking as only a millipede can, with its thousand heels. Deserting for a moment Opalina's point of view, my scientific conscience insists on asserting itself to the effect that no millipede with which I am acquainted has even half a thousand legs. But not to quibble over details, even a few hundred kicking legs must make quite a commotion in Opalina's home, before the pepsin puts a quietus on the unwilling invader.

From now on there is no lack of food, for at each sudden jerk of the whole amphibian there comes down some animal or other. The vegetarian tadpole with its enormously lengthened digestive apparatus, has crawled out on land, fasting while the miracle is being wrought with its plumbing, and when the readjustment is made to more easily assimilated animal food, and it has become a frog, it forgets all about leaves and algæ, and leaps after

JUNGLE DAYS

and captures almost any living creature which crosses its path and which is small enough to be engulfed.

With the refurnishing of her apartment and the sudden and complete change of diet, the exigencies of life are past for Opalina. She has now but to move blindly about, bathed in a stream of nutriment, and from time to time, nonchalantly to cut herself in twain. Only one other possibility awaits, that which occurred in the case of our Opalina. There comes a time when the sudden leap is not followed by an inrush of food, but by another leap and still another and finally a headlong dive, a splash and a rush of water, which, were protozoans given to reincarnated memory, might recall times long past. Suddenly came a violent spasm, then a terrible struggle, ending in a strange quiet: Opalina has become a link.

All motion is at an end, and instead of food comes compression, closer and closer shut the walls and soon they break down and a new fluid pours in. Opalina's cyst had dissolved readily in the tadpole's stomach, but her own body was able to withstand what all the food of tadpole and frog could not. If I had not wanted the painting of a vulture's head, little Opalina, together with the body of her life-long host, would have corroded and

A CHAIN OF JUNGLE LIFE

melted, and in the dark depths of the tropical waters her multitude of paddle-hairs, her more or fewer nuclei, all would have dissolved and been reabsorbed, to furnish their iota of energy to the swift silvery fish.

This flimsy little, sky-scraper castle of Jack's, built of isolated bricks of facts, gives a hint of the wonderland of correlation. Facts are necessary, but even a pack-rat can assemble a gallon of beans in a single night. To link facts together, to see them forming into a concrete whole; to make A fit into ARCH and ARCH into ARCHITECTURE, that is one great joy of life which, of all the links in my chain, only the Editor, You and I —the Mammals—can know.

II

MY JUNGLE TABLE

MANY, many, many years ago, in some distant place, among trees or rocks, perhaps on the banks of a river, certainly in the warm light of the sun, one of your ancestors and mine became tired of squatting on a branch or on the ground, and sat himself—or herself—on a fallen log. If it was himself then he must soon have felt the need of a lap on which to rest things—his hands if nothing else. And from that day to this, his male descendants still feel that lack down to the last unfortunate who is handed a cup of tea or a three-legged eggshell of cocoa, a serviette and a cake, with no support other than wholly inadequate knees.

Of the first table I can relate nothing with certainty, but of the last I could gossip endlessly, limited only by writer's cramp and my supply of adjectives. For I am at this moment sitting at the last table ever made—last because it is not quite finished. I am forever tacking on a little shelf or an annex at one side, and so I feel a right to place it at the opposite end of our distant forebear's piece

MY JUNGLE TABLE

of bark or stiff frond or whatever it was that he balanced on his hairy, bowed knees. And yet his table and mine are much more alike than the mahogany roll-top with swinging telephone and octave of assistants' push buttons to which our more sophisticated but less happy bank presidents sit down.

That reminds me, however, that my laboratory table is also of mahogany, because here in the jungle of British Guiana it is the cheapest material in the form of boards.

The crab-wood top grew in this very jungle, its first, rich red-brown cells fashioned from the water and earth and sun at least a century and a half ago. It is possible to detect the double character of the rings, indicating the two annual rainy seasons—the two springs which quickened the sap and leafage, and the two periods of drought when the life of the tree slowed down. Close to the heart of the great board is a strange ring, or rather node between rings—a wide, even space, which my reckoning places about 1776; about the time when our forefathers were fighting for freedom, whose memory we may not toast even in wine; they had just penned a Declaration of Independence, whereas we are considering passing a law to keep monkeys in their proper place. I pause in my table talk long enough to thank heaven that we are still allowed to believe in the rotundity of the earth, that the Indians' gift of tobacco is still permitted us, and that tea is not yet thrown overboard!

The year 1776 at Kartabo was one of almost continual rain,—so much my broad, crab-wood space shows—with no slack-growth period for this slender sapling. And imagination helps us still farther when we recall something of the human history of the place. Ever since 1600 the Dutch had strived to make this region habitable. The little fort on the island off shore had bravely pointed its guns down river, had fired its well-weathered cannon in victory, and had silenced them in defeat to English and French privateers (often an old-fashioned way of pronouncing pirate!). Hundreds of Indian slaves had worked on the four large plantations and only in 1772 had the settlers admitted that this region was fit only for the jungle, wild animals, and future enthusiastic scientists with tables. And now I realized that my table-top had sprouted in the very year that the Dutch left for the coast—one of the first wild things to spring up in their retreating footsteps, a pioneer in again "letting in the jungle."

The magic of my jungle table is always apparent in one way or another. No thoughts which it gen-

MY JUNGLE TABLE

erates, nor happenings on its surface are aught but vivid, vital, memorable: It is an event to hurry out to in early morning, it is a regret to leave for jungle tramps and for meals, it is only exhaustion which excuses its midnight abandonment. A magic carpet transports one's body from place to place, whereas my table impels mental gamuts from quiet meditation to dire tragedy, from righteous anger, to wonder at the marvellous sights it vouchsafes me, and despair at thought of their interpretation. Only once have I ever become impatient with my artificial lap, when an injury to my foot compelled me to remain indoors for a time. Then indeed the jungle called and *les affaires de ma table* palled,— a commentary on my lack of philosophy.

The first magic which my table made was to prove to be alive. The top was undeniably dead, well seasoned and inert, but my black boy Sam had cut the legs from jungle saplings. I put my hand down one day and felt a soft tissue something, half way to the floor. It seemed a moth's wings or a tangle of dense cobwebs, but to my surprise I saw that my table was sprouting leaves, rather pale and dwarfed, limp and flabby, to be sure, but of rapid growth, and besides there were four other buds just started. I had put cans of water on the floor beneath the legs to discourage ants, and the sap of the new-cut poles had greedily sucked this up, and even in the dimness of the laboratory light had begun to spread into foliage. It was proving a real jungle table and I was rather thrilled to see that the warfare of the wilderness had already begun at arm's reach,—a tiny caterpillar had crawled from somewhere to the new blown leaves and had eaten out a bit. I pictured my table as sprouting, growing higher and higher, until, in lieu of Alice's toadstool, I cut jungle saplings for my chair legs too, and mounted with the table! The Indian summer of my table legs soon passed however, the sap dried, the leaves wilted, and from saplings they became furniture.

But the magic continued. If the crab-wood boards of the top were not quickened into even passing vitality, they could do equally surprising things, the first of which was to become vocal. Day after day there arose a low grating throb, lasting for a few seconds, and sometimes increasing in rapidity and pitch until it assumed a true musical quality. Its direction eluded me until I happened to have my ear close to the table, when the vibrations seemed to sound at my very ear-drum. Then one day I noticed a tiny pile of sawdust on the floor and traced it to a rounded hole from which at intervals came the sound. For three months my

MY JUNGLE TABLE

musical table continued its monotone, day and night, until in the quiet of midnight it became part of the silence, and I was aware of it only with effort. Then it ceased, and its cessation held my attention more than its occurrence had done.

Months later when the last of my small table furnishings had been packed, I tipped up the table to carry it away, and there in the hole from which the monotone and the sawdust had flowed there hung suspended a gorgeous, mummified beetle, its long antennæ of salmon and black curved up and over its back, while its fluted cuirass shone through dust and dim light, deep forest-green framed with a delicate border of primuline yellow. My table top had furnished nourishment, sanctuary, sounding board, through all the long period of immaturity, but at the climax of this little life, the hardened vegetable fibre had held firm, despite all the efforts of the green beetle, and cruelly withheld freedom by some slight, needless entanglement of its hind legs. So passed two tragedies of my table,—the first vegetable, the second animal.

Usually my table is littered with beautiful mysterious things which, to a casual onlooker, could have absolutely no meaning. There is a small, exquisitely molded bony cup or vase, partly covered at the top, and with a long, daintily curved

JUNGLE DAYS

handle, which I keep suspended as a receptacle for pins. It might well be a delicate netsuke carved in pre-democratic Japan by some craftsman who wrought for love; it might be almost anything but a music-box. And now my reverie was interrupted by a sound from the neighboring jungle,—a sound common but never old. As the bony box might have been far other than it was, so the deep vibrations could well be elemental,—a distant wind, sinister as if it came straight from blowing across terrible fields after battle, or through cities wracked with pestilence; the caves around which it had howled must have been very evil, roofing ancient castles which sheltered thoughts of treachery and deeds of unfair violence. But I knew that the rich primeval resonances came echoing from bog bony boxes exactly like my pin holder, in the throats of a tree-top circle of beings like aged, thick-necked dwarfs squatting high on swaying branches, looking out toward me over the expanse of quicksilver water. At the climax, when it seemed impossible that any one animal could produce such an appalling volume of sound, a blur swiftly feathered the surface of the river, as if the impinging ululations of monkey voices had actually been translated into visibility—as liquid in a glass is troubled in sympathy with certain chords of music. My ear

MY JUNGLE TABLE

changed focus, and like a search-light shifting from distant cloud to airplane, attended a sound at my very elbow, throbbing, muffled—and again my table sang.

Amazing things, things apparently well within the realm of black magic occur and recur on my table. Late this evening a windless tropical rain fell so evenly and steadily that the monotone on the bamboos seemed intended for some other sense than the ear. I sat describing the delicate arrangement of the tiny bones and muscles of the syrinx of a flycatcher, striving to understand how there could emanate from this instrument such an intricate vocabulary of screams and whistles, trills and octaves as this bird and its fellows uttered every day in the laboratory compound.

Suddenly something flew swiftly past my face and alighted clumsily among my vials and instruments. I saw a giant wood roach all browns and greys, with marbled wings, strange as to pigment and size, but with the unmistakable head and poise and personality of a New York "Archie." The insect had flown through the rain and into the window, but a glance showed that it was in dire extremity, being in the grasp of a two-inch ctenid spider. The eight long legs held firmly, but had not been able to prevent the roach from flying.

JUNGLE DAYS

At the moment of alighting the arachnid shifted its grip, and secured the wings so that further escape was impossible. Both were desirable specimens and I instantly slipped a deep stender dish over them and again lost myself in my binocular microscope.

Fifteen minutes later I looked up and saw a sight so strange that Sime himself would hesitate to delineate it. The spider still clung tenaciously to its victim, but the wood roach had her revenge. She was barely alive, yet in a quarter of an hour she had changed from a strong, virile creature to an empty husk, dry and hollow, while over her and the spider, over glass and table-top scurried fifty active roachlets. They had burst from their mother fully equipped and ready for life, leaving her but a vacant, gaping shell, a maternal film, the ghost of a roach: Tiny, green, transparent, fleet, they raced back and forth over the spider. He grasped in vain at their diminutive forms at the same time still clutching the dying, flavorless shred of a mother roach, holding fast as though he hoped that this unnatural miracle might reverse itself at any moment, and his victim again become fat and toothsome.

I knew that some of the fish swimming in the aquarium near by lay thousands of eggs, and that

"Well within the realm of black magic"

"Silent and smooth as a mirror"

MY JUNGLE TABLE

other insects leave myriads of offspring, yet this magic of the wood roach, this resolution of one into fifty made wonderfully vivid the reproductive powers of tropical creatures. When in a moment of time, relatively speaking, a single insect can be broken up into half a hundred active, functioning duplicates of herself, the chance for variation, for new adjustments, for survival of the more delicately adapted is faintly understood. Here was spontaneous generation with a vengeance.

To hark back again to sounds and voices; I could fashion a whole essay on the calls and songs and noises which come to me at my table, from river, compound and jungle. On very still days I can hear the giant catfish thrumming deep beneath the water, and the cry of hawk-eagles high in the heavens; at hot, high noon Attila, the brain-fever Cotinga, calls and calls and calls, while through the hush of midnight there comes the hopeless cadence of the poor-me-one; I know from a sudden babel of humming-bird squeaks and frenzied shrieks of flycatchers that a tree snake has been discovered in the bamboos; I am certain without looking that it is very close to five o'clock, when the first old witch cuckoo begins whaleeping on its regular evening excursion for a drink in the river, and so on.

JUNGLE DAYS

Probably by virtue of my table's magic, I have learned, like Chubu and Sheemish, to work a little miracle all by myself. My principal technical work just now is the study of the syrinx of birds, their remarkable, complex organ of voice placed far down beyond the throat, in the very body itself, and the correlation of its structure with the actual voice of the bird. At present I try to solve some knotty problems of tinamous, strange, bob-tailed game-birds, related both to fowls and to ostriches, which live on the jungle floor, lay eggs like burnished turquoise and age-purpled jade, and call to one another with sweet, liquid whistles. My Indians bring in numbers of these birds for the mess, so I have an abundance of material for study. I try an experiment on my table, which has already been successful in other cases. I decapitate a bird before it is plucked for the pot, and holding it firmly on its back, I strike a sharp blow on the muscles of the breast. Nothing results, so I shift position and try again. This time a short, high note is produced. I draw out the neck a little and obtain a lower note, still further and strike a half tone lower in the scale. If I could prolong these I could reconstruct the whole plaintive evening call of the variegated tinamou here on my very table top.

MY JUNGLE TABLE

Then I take the windpipe and carefully work out the wonderful architecture of the whole organ, the delicate adaptation and adjustment of each part fulfilling its special function, the whole working together as no man-made machine ever could. From throat to syrinx the windpipe extends, composed of thin membranous tissue, kept open by a series of a hundred and twenty-five perfect rings. Here we have assurance of an entrance for air forever clear and open, so mobile that it bends back double, yet with no chance of closure through any contortion of the neck. The throat end is guarded by a slit which opens and closes at the slightest need; the opposite end marks the top of the syrinx and the division into two tubes, each leading to a lung. For twenty rings above this point, the windpipe is slightly enlarged and almost solid, forming a bony sounding board, which acts, in a less degree, like the throat box of the red howling monkeys, giving resonance and carrying power to the voice.

The syrinx itself is boxed in by four pairs of large rings and semi-rings, which protect two pairs of cartilage pads. The pads of each pair touch one another along their inner sides, and when the windpipe is relaxed the seam between them is closed tight. A slight tug, as in my decapitated bird, corresponding to a raising of the head and neck in

JUNGLE DAYS

a live individual, and the pads revolve slightly, bringing a constricted part of each into the seam, forming a tiny gap. Through this the air from the lungs and air-sacs rushes and we have the mechanism of the first, high, clear note of the call, a superlatively sweet whistle on middle C, carrying a mile through the thick jungle. Although quite another story, my mind rushes on, away from the technical anatomical problem, to the realization that this sound is a summons from the very advanced female of this species to any unattached male bird, an announcement that she is ready to lay an egg for him, provided he will incubate it, hatch it and assume entire charge of the young bird. And I do not know whether to cheer or blush for my sex when I state that the woods hereabouts are full of amiable, domestically inclined males who are eager and willing to agree to this rather one-sided contract. Their syringes are almost identical but the loud evening calls are invariably those of the idler sex. Notes for Women! must have been the slogan of the long since successful tinamou suffragists.

It is amusing to trace a circular gamut of human interest in animal sounds: Listening to various screams, warbles, whistles, roars, chirps, trills and twitters in the jungle, an intelligent interest impels

MY JUNGLE TABLE

us to desire to know the author; having accomplished this by patient stalking and watching, and if needs be, shooting, the wish is aroused to discover the accompanying emotion, the incentive, and then the fascinating problem presents itself of the answer, whether in terms of action or vocal, whether filial, amorous, pugnacious, or merely companionable. This is more difficult, but in many cases possible. Almost always this ends the quest, while it is still incomplete. The method, the physical mechanism, is after all, the foundation of the phenomenon, and when we have secured a specimen, taken it to our table,—a tinamou in the present instance—then we may produce the call artificially, and by tireless and detailed dissection detect air channel, resonance chamber, syrinx mechanism, vocal chords, controlling muscles, and envy the enormous bodily reservoir of air—lungs, sacs, the very hollow bones themselves. Leaning back and listening to a living, wild tinamou calling in the neighboring forest, feeling rich in the possession of its Who! Why! and How! we realize the fullest joy of intimacy with the furtive beings of earth, with the elusive small folk of the jungle.

After a long jungle tramp I was leaving Hacka Trail for the Station clearing, when I caught sight of a group of small objects on the under side of a gigantic bromeliad leaf. If the leaf had been fifty feet up they might have been great fruit bats, if twenty feet their size would have equalled that of vampires, but as they were only out of arm's reach above my head they could not be more than an inch in length. When I had hacked off the leaf and dodged its fall, I found nine little chrysalids clustered together, and even on close scrutiny their resemblance to a group of diminutive bats was still absurdly real. This intimate association of chrysalids is a rare thing, as rare as the nocturnal association of butterflies sleeping in jungle glades.

I carried off the leaf curved into a great emerald arch, and fastened it over my table, where it dried into a fluted dome of green tissue. Three days passed with no sign of change from the chrysalids swinging from their silken pendants, when my eye caught a glint of silver far down the under side of this same leaf, near the tip. Another glance made me think them inexplicable dewdrops, a third crystalized them into pearl-like consistency, while a fourth careful scrutiny showed me they were two eggs of a scarlet and black heliconid butterfly, the kind which fluttered fearlessly ahead of me along the jungle paths. Here was a splendid example of oblique discovery, of scientific second sight.

I wondered what sculpture the surface would

MY JUNGLE TABLE

show,—these two isolated spheres, shining like the third zodiacal sign against a dark green heaven. At the first look through the microscope I forgot all about surface and possible spines or hexagonal lattice-work; it was the contents which drew and held my attention. A butterfly egg in due course of time should yield a caterpillar, which before it emerges is wound into a curve to fit its minute spherical home. But here was a new cosmos,—a planetful of slowly moving creatures which had nothing in common with a heliconid caterpillar. Slowly they milled around their little world, living, like some Gulliverian organisms, on the inside looking out. The egg was an opalescent sphere, a twelfth of an inch across, and in my microscope field it seemed really suspended in space,—in a dark chlorophyll ether. More than once as my eye tired in watching I seemed to see the whole egg revolving while the inmates remained stationary. Now and then one of the egg-beings turned and went against the current, setting up a traffic whirlpool which caused all to cross and recross in confusion. The film of eggshell was translucent and clear immediately beneath my eye, clouding into exquisite purplish pearl at the periphery. One of the inmates came to rest directly beneath the surface, and I saw it was a tiny grub, legless, searching about blindly, feeling,

JUNGLE DAYS

sensing, living, after whatsoever manner grubs live who find themselves prisoned in a butterfly egg. The grub hastened on, fell into wriggle with its companions and soon slipped from view below the edge of its world. Doubtless in a few seconds it completed its internal orbit and again crossed my field of view, but like a circulating Roman army on the stage, or the sequence of ideas in some sphere not attached to jungle leaves, all seemed identical. I could never tell when the same one appeared again; indeed while they moved I could make no estimate even of their numbers. I only knew that some minute hymenopteron, doubtless a member of the wonderful tribe of Chalcids, had, a few days before, thrust her ovipositor through this translucent pearl and left within as many eggs as there now were grubs, then flown on to the next egg. I once was fortunate enough to observe this fairy egg-laying,[1] and now I was trembling with excitement at the unexpected treasure trove I had unwittingly brought to my table.

Closest examination from every side with high-power lens revealed to me no hint of the place of entrance. Once when I crawled from the heart of great Cheops out through the robbers' tunnel, and finally scraped and squeezed through the narrow

[1] *Edge of the Jungle,* pp. 38-40.

MY JUNGLE TABLE

crevice through which they had broken in, I thought it small indeed. But here was a phenomenon far more wonderful than a full-rigged ship in a bottle, a snow-storm in a paper weight, or the thieving Arabs' entrance in the pyramid.

Four days passed, the wonderful globes lay before me, and then I examined them again. A remarkable change had been wrought, a living planet had devolved into a dead satellite; the egg had become a sarcophagus with a dozen mummies. The little cases were arranged around a central core of débris, some standing on end as in the Egyptian room of a museum, a group facing one another as some wordless gossip passed from one sealed mouth to the next. A single mummy doll rested against the opal shell, with eyes pressed close to the translucent pane, eyes which at present existed only in outward form as insensitive tissue. This one individual had chosen for his final pupal change a position at the very outer rim, where the first nerve tingles of sight would reflect the mysteries of the world beyond that sphere of food and fellows which had heretofore bounded his existence; my masculine pronouns are adumbrative, not casual.

So passed a week with the little silent mummies still unchanged; seven days,—sufficient time, Biblically speaking, for the creation of the world.

But just as all the glorious truth and beauty of evolution is concealed within the metaphor of Genesis, so, hidden from our groping senses, miracles of change were being wrought within the butterfly's egg. The following morning the spell had broken, and the sphere again seethed with life, resurrected, reincarnated. On the central compost heap were piled twelve suits of second-hand pupal skins, tissue paper cartoons of their wearers, glimmering weirdly through the shell. The tiny wasps had all emerged and were active, and already there was a hole bitten through, with small ships of splintered opal scattered outside. As I watched, a wasp midget shoved aside a group of idlers, pushed his way to the door and began to gnaw with all his might. His great bulging scarlet eyes blocked the way as he tried time after time to press through. The whole eggful knew that something of great import was happening, and the outside air must have carried exciting tidings, for all moved about as quickly as their crowded quarters permitted. Twice the Gnawer left his labors and walked about nervously, once making the entire circuit of the egg. His leadership, his pioneer daring was marked not only by action; I found that I could readily distinguish him from the others. He was a shade smaller, his lines were trimmer, and

MY JUNGLE TABLE

upon his back was a round insignium of gold which the others lacked.

Several others came to the opening, tried to pass and turned aside—none made attempt to aid in the escape from prison. Back came the ambitious one and fell to with all his strength. He lacked leverage, and only when three of his companions came up at once, was he able, by pressing his hind legs against their faces and bodies, to break off an unusually large bit of the horny shell. This made a splendid gap, and after two smaller bits had been chewed off, the little insect wriggled through the jagged hole, and stood upon the summit of his world. Tiny though he was, needing thirty-five of him to cover an inch of space, his coloring was exquisite; eyes dull scarlet, sparsely covered with golden hair, body armor of glistening black from head to tip of abdomen, with badge of yellow gold shining from between his wings. These wings were small, paddle-shaped and almost free of veining, while the scales on their surface glowed with iridescent play of lilac, yellow and pale green.

Now ensued an elaborate cleaning of every part of his body, and then he ran off at top speed. Several quick turns near-by on the leaf and back he came, gave a final wipe to his forelegs, climbed up, antennæd the hole and took his stand a wasp's

JUNGLE DAYS

length away. This action came as a complete surprise; I never expected him to return after such a laborious escape.

Soon a second wasp came to the breach and squeezed through. Hardly had its combing and scraping been completed when, to my astonishment, the Gnawer rushed forward, roughly seized the second wasp and began to bang its head most unmercifully. At every push, the head of the unfortunate insect wobbled as if about to fall off. Suddenly it rose to its feet and the first wasp mated with it. I then realized that instead of assault and battery, this was courtship, that in place of horrible fratricide, this was the nuptials of brother and sister. The mating lasted but a second, when the first wasp returned to its watchful waiting, and the other spun its paddle-shaped wings and flew off as far as the confines of the covered glass dish permitted. I never took my eye from the lens as the miracle continued. One after another the sister wasps emerged, to the number of eleven, and in each case the male enacted his rough courtship and mated for not longer than two seconds. In each case, without a moment's hesitation, the female flew swiftly away. Once, when three emerged quickly one after the other, they did not leave the egg but waited quietly for the male.

MY JUNGLE TABLE

The whole thing began and ended so quickly that it was some time before I could review the whole wonderful performance from the conjectured laying of the eggs, through the grub, pupa and now the adult stage. I looked again at these midgets, only a thirty-fifth of an inch in length, and considered their necessities in life,—food, mate and a butterfly's egg, and I realized the enormous advantage of this simplification of the mating problem. But the most astonishing thing of all was the thought of the anticipation, of the perfect adjustment of sex in the unformed organisms, the pre-natal compulsory affiancing, together with the apparently satisfactory disregard of inbreeding adumbrated in the very eggs themselves of the original mother wasplet.

No matter how imperfectly I have translated this event, disregarding my futile phrases and in spite of my inadequate description, it was a most wonderful happening, which for a time completely eclipsed all other affairs of my table top. In delicate achievement, astounding unexpectedness and magical matter-of-factness, it left the onlooker with a supreme realization of ignorance and a dominant sense of awe.

And so as I sit at my table, my little cosmos of space and time presents deaths by violence, and

JUNGLE DAYS

lives of quiet, unperturbed peace; acrid, burning odors and smashing, sweeping brilliancy of color; living skin soft and smooth as clay, or fretted like shagreen; voices almost high enough to become visible; comedy so delicate that appreciation never reaches laughter, and tragedy so cruel and needless that it stirs doubts of the very roots of things. All these and many more, begin, occur and pass before me,—things which go to make up a world.

III

A MIDNIGHT BEACH COMBING

A TROPICAL night may be quiet and calm, and yet full of a strange restlessness. It was such a one when I lay in my bathing suit close to the grey granite of Boom-boom Point, and watched the low-hung North Star twinkling through the fretwork of mangrove roots. Three great planets added their separate lustre, Mars overhead in the very heart of Scorpio, Jupiter well down to the west, and Venus just setting, shining with the light of a half moon. It was, however, predominantly, a Night of the Milky Way. The great luminous highway stretched from horizon to horizon, illuminating hundreds of the tiny mica facets in my rocky couch. Great Cygnus climbed slowly, majestically, along the glowing path, and Pegasus reared his head just above the horizon. Has the composite light of these myriad stars the same sinister psychic effect as the moon rays? Else why were I and so many creatures restless? Only the giant tree-frogs, the Maximas, wahrooked in endless, stoical reiteration, unaffected by stars or planets, as endless as an after-dinner speech and as unintelligible. Now and then a trio of Typhon's toads exploded in a short, hysterical outburst, as if intercalating *Hear! Hear!* or *Cut it out!*—a very impudent, undesirable, nervous protest against the brain-fever repetitions of the great frogs.

I was ready for something unusual, and it came, —merely a sound, but one which will probably be as mysterious on the day of my death as it is now. Without warning, through the air overhead, against the translucent celestial glow, came an *izzzzzzzz-wonk! wonk! wonk!* as evanescent as the low twang of a bullet, wholly indescribable in its true weirdness and richness. No beetle ever turned as quickly as the *wonk! wonk! wonk!* indicated; no bat ever achieved a twang with its velvet wings. It was no sound of bird or insect that I knew; and it came again and again from the same direction, and seemed to emanate from some creature which watched me. The *wonk! wonk!* as of sudden, banking flight, happened close in front, over the water. I flashed my electric torch and saw nothing, although the sound continued, and for half an hour one or more mysterious beings swept about me close overhead. As once before, my mind went to Pterodactyls and I imagined a pair of the little

A MIDNIGHT BEACH COMBING

web-fingered creatures launched out from some secret crevice in the distant mountains, for a brief time to hawk about in the light of the Milky Way, peering down with their great eyes, toothed beaks half open, whipping back and forth through the air, now and then snapping up a bat, and stirring the imagination of a curiosity-tortured human, who would willingly give a year of his life to see such a sight.

I had meant to spend part of the night among the mangroves, but the glimmer of the white sand drew me up instead of down the shore, and I crept over the rocks and padded silently over the sand to our swimming beach.

The tide was half-way down, silent and smooth as a mirror with every star doubled. As I watched, they were erased, one by one as if the reflections had become water-logged and sunk, and looking up I saw a mist swept by the high trade-winds across the sky, while around me not a breath of air stirred. I wriggled into a form half below the surface of the sand; I worked down lower and lower until I was at the very edge of the water, which is one of the most wonderful spots in the world. Being there is the very least part of it. Thousands of people are there all through the summer at Coney Island and Margate, but never think themselves anywhere but swimming at Coney Island or bathing at Margate.

Between tides is really the wildest place left in the world, the truest no-man's-land, for while you may sail in all waters just beyond or loll in a hotel a few yards behind, you cannot remain where you are except anchored and in a diver's suit. And whatever man erects there is sooner or later smashed into joyful chunks of cement by the storm waves. The delight of it is to feel yourself as I did at this moment, a third under water, a third buried in solid sand, and the rest of me bathed in and breathing the air. We sometimes feel a thrill at bestriding the border line of two states or countries. How tremendously more wonderful to snuggle close to the three states of matter, solid, liquid and gaseous, and then indeed to realize it and thrill to it with what seems a fourth state—the mental and spiritual.

The crunch of the sand grains, the lap of the water, the breath of air,—it makes the world very primitive and new. Without my flash I can detect no hint either of vegetable or animal kingdom—my little cosmos at the meeting place of the elements is wholly inorganic, and mind. If only earth-fire were added, it would be complete, and here, a hundred feet from my cot, there would truly be an epitome

A MIDNIGHT BEACH COMBING

of the primeval earth. I wonder however, whether it is all not more adumbrative of ages to come, when the last animal has fallen, the last leaf shrivelled, and only the inorganic and spirit remain, than of the infinite past.

My day-dreams or rather nocturnal meditations were leading me into hypnotic depths when, with a single bound, I deserted my most ancient medium, water. Momentarily I even left my more recently ancestral acquisition, earth, and entered the third which I had conquered only during the last eight years. Gravitation, faithful through all physical and mental vicissitudes, brought me down with a resounding thump. At first I was simply dazed. What had happened? From the infinite calm of abstract meditation I had been galvanized into the most violent paroxysm, and here I was sitting on the sand, unhurt, stupidly wide awake, with my heart trip-hammering. Then all at once the physical me calmed down and the mental took charge, first in a thrill of excitement at realization of what had happened, then in joyous recognition that, as at a well-planned dramatic dénouement of a play, the miracle had happened. Nature, tired of being ignored, had entered my inorganic make-believe cosmos, completed it and split it apart with a vengeance. Instead of sending a firefly into my ken, she had been more subtle, and an electric eel had brushed against the sole of my foot, and discharged his diminutive broadside. The shock had been slight, but unprepared as I was and completely relaxed, it had seemed to my nerves like the discharge from a third rail. With my flash I caught a momentary glimpse of the lithe black chap, and I dabbled my hand in his direction, but he eeled away and became one with the dark water.

I could not get back to my former isolation, even if I greatly desired to do so; the eel had changed all that. He seemed so modern, so conventional and specialized an organism, drawing the lightning down into the dark waters, and liberating it at the will of his fishy brain.

I rolled over and flattened myself, and with my electric torch held at eye height, horizontally, I entered one of the strangest of worlds,—a beach at black midnight. My mind kept wandering back to my trio of elements, and I thought of the water ouzel which has conquered them all. In the wilderness of western China I have seen this delicate, thrush-like bird run rapidly in and out of a tangle, over leaves and sand to the edge of a high river bank, and then taking wing, fly in and out between the boulders of the stream, finally to dive headlong into the swift water and creep along the bottom,

A MIDNIGHT BEACH COMBING

feeding as it went. Here, in the space of a minute or two, was exhibited mastery of earth, air and water; only the phœnix could claim superiority.

This evening I was to find a living rival to the ouzel, an insect, a cricket, which, like so many wonders, was not in the heart of the Asiatic continent, but at the very door of my British Guiana laboratory. In the level glare of my flash all the beach creatures became unreal and of low visibility, while their shadows took full possession. This fanciful phrase reflected a very real and interesting scientific fact, that the reason for this lay, not in the unusual lighting, as much as in the color of the little people themselves. Picking its way over the sand came a low-swung, weird, blackish thing, whose silhouetted head swung from side to side, and just above it there appeared a fearful thing, on long emaciated legs, which crept nearer and nearer, and finally rushed at the first and sank down upon it. The attack was so sudden and the images relatively so huge that I involuntarily sat up and raised my light. The two images rushed toward me and vanished and my eyes suddenly shifted to nearer focus. I had been watching the shadows of a small insect and a daddy-long-legs, the substance of which now appeared ridiculously small and close to me, with their shadows well under control beneath them. Slowly I lowered the flash again, and in spite of all I could do, my eyes gradually lost the creatures themselves and followed back along the lengthening lines of legs, to the gargoylesque false phantoms,—the gyrating monstrous phantasmagoria on the sands. Never have I seen a more completely sense-deceiving phenomenon. Sitting up, I looked down upon small, slowly moving, barely distinguishable beach beings; prone, I was surrounded by unnamable apparently ectoplasmic ghosts. If I should accurately describe their anatomy and actions as revealed by my low-hung light they would fit into no living or fossil phylum of earthly organisms. By shifting back and forth I again focussed on the terrible battle going on at my side, and now the giant had lifted the lesser beast bodily in its jaws, and was staggering about, mumbling it as it went. My scientific terms locustid and phalangid faded from mind with their substance, and I lay watching the midnight shadow struggle between Plash-goo and Lrippity Kang.

I had always thought of daddy-long-legs as harmless living skeletons, who clambered aimlessly about and dropped their legs at a touch. Now I found that they could be ravenous beasts, their dwarfed and rounded body swung high aloft on their eight thready legs, creeping over the sand,

A MIDNIGHT BEACH COMBING

and actually running down, pouncing on and killing insects as large as themselves. In this case it was a green grasshopper nymph who was seized, bitten and worried with an unnecessary amount of dragging about and vicious chewing. I leaned slowly forward with my hand lens until I could see every detail, and if daddy-long-legs were magnified in life only fifteen times I should flee in terror from what would be a worse danger than any wolf. The horrid eyes, grouped in their solid clump seemed to be even now watching me malignantly, and the great needle-sharp fangs were sunk deep in the grasshopper, and being worked back and forth as the juices of the still living insect were sucked up.

Soon the creature set to work to sever the abdomen from the rest of the insect, and the head and legs fell to the sand, the feet waving slowly and vaguely. The daddy-long-legs did not move, except now and then to lift one or two legs and hold them aloft when a passing ant brushed against them; twenty minutes later it was still there, draining the last drop from the shrivelled grasshopper.

My attention was attracted to the approaching shadow of another spectre, only in this case the shadow was indefinite, humped; it might have enshrouded a low fluttering moth or awkward beetle. Instead of which, when I followed down the shadow

JUNGLE DAYS

path to its substance there loomed suddenly a figure even more terrifying than the daddy-longlegs. But this was awful in a wholesome way. You started at first sight, then smiled, then felt a liking for the apparition. It was decidedly the Personality of the beach, claiming full attention as long as it was in sight, clownlike in its comicality, and childlike in its seriousness and the affection it aroused. Many will doubtless wonder mildly at thought of the possibility of holding a mole cricket in affection or esteem. Yet it is true that when I return in memory to Kartabo, my thoughts of beauty go to the great blue Morpho butterflies, of grace to the soaring vulture, of adorableness to infant sloths, and of amusement and affection to the jolly white mole crickets of the sand.

These are the chaps who fairly outdo the water ouzel, outflying, outrunning and outswimming that bird, and in addition being powerful leapers and the most perfect burrowing machines in the world. Unlike their neighboring relations of the jungle these shore crickets have taken on the color of the sand, keeping only a few hieroglyphics of dark pigment. Their eyes alone remain solid black. No matter how deserted the beach, how lifeless the tropical jungle may seem, I was always certain of finding these optimists abroad after dark, scurry-

A MIDNIGHT BEACH COMBING

ing here and there, or popping unexpectedly up from the wet sand which a few minutes before had been covered with the tide.

As my new visitor approached, after my first emotion I was able to call him by name, a name as bristling with sharp-angled syllables as the tips of his front legs. Indeed his sponsors must have been profoundly impressed with these great limbs for in *Scapteriscus oxydactylus* they dubbed him the Shovel-winged, Sharp-fingered One.

In the month of March I found little spurts of wet sand on the upper beach, and following down each tiny hole for an inch, I surprised a diminutive white cricket, almost a replica of the large ones, just hatched and bravely starting out in life for itself. In the following months their numbers sadly diminished and the size of the few remaining individuals increased, being gaugeable exactly by the calibre of their hole which they open when the tide goes down. Now, later in the year, the adult mole crickets were in the full prime of life, vital, virile, meeting on equal terms all the dangers and advantages of nocturnal life on a tropical beach. I appreciated these insects all the more because of their local distribution, being found nowhere up or down the river, except on our short stretch of sandy beach.

JUNGLE DAYS

The hind legs are swollen with muscles for leaping, and with broad, flat soles for pushing, the middle legs are normal supports, but the front ones are a study as scientific, mechanically perfect excavators. There are sharp, horny, downward-projecting pickaxes, lighter pitchforks, backed by spade-shaped implements, and bordered with stiff, broom-straw edges for sweeping away the loose débris. In fact this little insect has everything but dynamite for making easy its passage underground. It even has long feelers behind as well as in front of the body.

Like the kick-off of a big football game, or Fred Stone, or a shark on your fish line, when one of my mole crickets came into sight, I knew that something exciting was certain to follow. On this midnight, while the big insect had zigzagged toward me, the tide undermined my sandy elbow-rest, and I slipped. At the first scrape of sand, he put both oxydactyl hands together over his head and half buried himself with three flicks. But he was neither coward nor ostrich and after a moment he had turned and rested his great arms upon the mound of sand, the strangest parody upon Raphael's cherubs imaginable. His head turned from side to side as he watched, and, I almost added, listened, for the source of danger. I remembered in time

A MIDNIGHT BEACH COMBING

that his ears were on his front arms just below the elbows, sandwiched between the pitchfork and the shovel. He twisted sharply to the left at the same instant that a miniature hidden mine was sprung, and a spray of sand shot upward. Almost before my eye could follow, a second mole cricket appeared, and each saw in the other the summation of all past troubles and future hatreds; they hesitated not a second, but flew at each other.

At first there was considerable side-stepping and feinting, and they whirled about one another until a well-marked ring was worn in the damp sand. Then they clinched and to my horror a leg flew up and off into the darkness. Now the timeworn, and at best inadvisable simile was reversed, and ploughshares as well as shovels, brooms, scissors and pitchforks were in a twinkling transformed into slap-sticks, swords, pikes and daggers. Twice the insects reared up on their hind legs, their arms working like flails. Now and then the lace-like wings unrolled and shot out as balancers, glistening like metal in the light of my flash. One cricket fell for a moment, the other pounced and a whole front arm rolled away. Nothing daunted, and indeed apparently lightened by the loss of his left arm, my cricket leaped at the other and bowled him over. I cheered—they both reared again—

JUNGLE DAYS

and were washed away in a tiny swirl of water,—the tide had turned and the first of the trios of incoming wavelets had caught all of us unawares. *Le duel nocturne des courtilières* was over. Each opponent had lost a leg, yet they scampered off and dug in with little appearance of crippling,—one limped a bit and the other sank his well somewhat obliquely, that was all. I remembered my first experience with these crickets, when I confined four together in a glass dish, and next morning found but one, large, plump and happy, surrounded with the crumbs of eighteen limbs; and I recalled the diminution in numbers of the broods of infant crickets, and I wondered whether I had better not slur over part of the home life of my little friends if I wished the mirror of my affection to remain untarnished.

I turned my light toward the water which was lapping shoreward, and on the surface were two white spots, mole crickets again, scurrying here and there with short strokes of the forearms, which had now become efficient oars. They soon sculled to shore and vanished, and a threat of moralizing came into my mind; how wonderful it would be if any of us could so completely master the conditions of life in our environment! Here were two sandy depressions where the crickets had disappeared; in

A MIDNIGHT BEACH COMBING

a few minutes the tide would cover them, and for eight hours thereafter the two bundles of vitality would remain buried beneath the waves, able somehow to breathe and to resurrect, to scamper about on their business of life on what remained of their legs, to spread their wings and fly wherever they wished—one place at least being to the lighted lamp on my laboratory table.

The wash of the tide made me restless and I swept my flash about in a last survey, when I saw a multitude of little orange-red lamps drifting toward me. Holding the light obliquely I saw the wraiths of many shrimps with their periscope eyes illumined by my electric wire. They swam steadily ahead, half blinded by the glare, until suddenly there came Nemesis with a rush and a swirl. I caught sight of long waving tentacles, a gaping mouth, flash after flash of glittering silver, and there at my feet was a catfish, half stranded with its headlong rush. Mindful of poisonous spines I flicked him up the beach with a hand blanket of sand, where he lay protesting with rasping twitters and peevish grunts until I salvaged him.

My last glance at the beach showed something so strange that I turned back, and discovered a wholly new field for enthusiasm. Many years ago I found that tracks in the snow could best be observed and photographed in slanting rays of the sun, and now my final, casual sweep threw out into strong relief a series of rabbit tracks; this in spite of the fact that I was some two thousand miles from the nearest bunny. Looking down at the tracks they completely vanished, not a depression or marking could be detected, but oblique lighting showed the scar of claw marks, all four feet close together, with a good eighteen inches between leaps. I puzzled long over it, I traced it almost to the water and up to the soft, dry sand. At last a thought came to me, and I went up to where I knew there would be, day or night, a file of leaf-cutting ants. There solemnly watching, and waiting for some favorable omen to begin her midnight supper, squatted my pseudo-rabbit, a huge, friendly grandmother of a toad. She blinked, and I reached down and tickled her side, whereat she grunted and puffed out prodigiously.

At this moment my eye wandered to a near-by bush and I made a discovery which whole hours and half days of intensive search and watch had up to this time failed to reveal. The line of leaf-cutting Atta ants led up this low shrub and many scores were deployed over the leaves busy on their eternal work of cutting off circular pieces. For years I had watched them carry these leaves back,

A MIDNIGHT BEACH COMBING

and had seen the free rides which many small individual ants took back to the nest on these wavering bits of leaf. Here, in the light of my flash, a medium-sized ant staggered along beneath a load, as if a man should balance a barn door on edge on his head. Like small boys hitching on behind a wagon, there were seven small ants clinging to the top and sides of the bit of leaf, probably doubling the weight, and altering the whole centre of gravity. I have seen a Japanese acrobat in the circus balancing a ladder with several men clinging to it, but this feat was infinitely more difficult. And there was no display to this. It was all in the night's work. These ants know not the meaning of play or vacations or any moment of unnecessary rest, and yet here were seven of them for their own convenience making much more difficult the labor of their larger brother, or rather sister. I knew there was some vital reason, some *quid pro quo,* but hitherto I had been able only to guess at it.

The small bush made all clear. There were enemy ants in the bush, who were attempting to drive away the Attas, and their scouts made attack after attack on the busy harvesters. Unless actually attacked and bitten, the Atta workers paid no attention to their assailants. I saw one partly crippled and yet go on with his load as best he could, playing pacifist for duty's sake. Their work was definite and inviolable, to cut a leaf and to transport it to the nest. The huge Atta soldiers, fat and enormous, who guard the depths of the nest and occasionally wander aimlessly along the line of march, getting in the way of their fellows, were nowhere to be seen, but the battalions of the Minims were in full action. They were too small to cut leaves or carry them, and had not even strength enough to walk both ways, to and from the nest. But on the leaves, facing the legions of the giant tree ants, they showed their worth, their *raison d'être*. I have never seen such fighters. They equalled the army ants, and lost leg after leg, even the whole abdomen, without slacking their efforts in the least.

On one leaf I saw a most exciting engagement. Three workers were cutting along the edge near the tip, and five small Minims were standing about with jaws raised suspiciously, when three black tree ants came on at once. One got past on the under side, tackled a worker and was seized in turn by one of the tiny bulldogs. The black ant let go the worker and tried to get at his tormentor, who had a good grip on his tender antenna. Chop went a leg of the Atta, but then another came to the rescue and got his jaws in a crevice of the armor

A MIDNIGHT BEACH COMBING

beneath the black body. This was too much and the trio fell from the leaf, out of the range of my light, into the darkness of the sand below. There were left three Minims and two black ants, the latter four times their size, and yet so furiously did the little chaps wage battle that the invaders had no chance to get past to the workers at the leaf edge. Another black ant now appeared, but close on his heels six Minims, and in the face of this squad they all fled minus a leg or two, and carrying three Minims with them who refused to let go, one of which had little of him left but his jaws which still retained their grip.

I saw only two workers killed or forced to drop their loads in spite of all the black tree ants could do. All the time new contingents of Minims were arriving, and in the midst of the hardest fighting, a little warrior would now and then climb upon a passing leaf and settle down for a rough trip home. It was as if they belonged to some autocratic labor union and had to punch a time clock at the nest, regardless of how things were going in the front line trenches. So the Mediums are the workers, the providers. The Maxims are the home guard, and the Minims are the standing army for border warfare, trudging bravely as far as they are needed to convoy the outgoing workers, but after battle or

JUNGLE DAYS

their share of watchful waiting getting a free ride home on any passing chlorophyll lorry.

Immensely pleased with the discovery of another detail of the Attas' life history I returned to my search for more sand tracks. Walking along the reeds with light held low, I saw clearly where an opossum had come out shortly before, dug a little in the sand and passed on, and most amusing was the record, in an isolated patch of clear, soft sand, of where a young one had fallen from her back, and straightway clambered on again. Farther on a big lizard had shuffled along, but the next track took me thousands of miles northward to New England sands in autumn,—the fairy footwork of a pair of spotted sandpipers which that evening, had teetered along the edge of this tropical river.

One last thrill my beach gave when, drawn by some instinct, I scanned the sand just beyond a clump of sedge. There, fresh and strongly etched, was a broad, sinuous line up from the water's edge, flanked alternately by crescents, deep bitten into the wet surface. This had been made by no creature with legs, but by some long, heavy body, alternately pushed up the beach,—the line and crescent sand signet of a great anaconda—king of all these waters, who, while I watched shadows a few feet away, had slowly drawn his mighty length past me,

A MIDNIGHT BEACH COMBING

up into the gully beyond,—who shall say where or why!

No wonder this night, so calm and peaceful on the surface had aroused an ill-defined suspicion of hidden things far otherwise. I looked out over the water, again alight with reversed constellations, I listened to the soft lapping of the rising tide, felt the first faint breath of the new day, and thought of the tragedies I had witnessed—the mole crickets nursing their wounds in their dugouts deep beneath sand and water, of the dead grasshopper nymph, the shrimp, the fire in whose orange eyes was forever quenched, and of the death struggles of the ants going on in the darkness at my feet.

The opossum was searching for food for itself and its young, and somewhere the great snake was coiled, watching with lidless, untiring eyes for its share in some life of lesser strength. It seemed somehow so cruel, this eternal alternation of life and death. If only the lower animals,—and then I remembered that perhaps at this very moment my Indian hunter was pulling trigger on an unsuspecting agouti or curassow or peccary for my next dinner; it came to me that the very emotions of compassion and sympathy which moved me, were materialized and sustained by the strength derived from the sacrifice of many, many lives of these same lower animals. I stopped thinking, stepped carefully over the line of insanely industrious Attas, and went to my hammock.

IV

FALLING LEAVES

NEXT to the dynamic crashing syncopation of a regimental band, or the subtle, infinitely more emotionally hypnotic beat of a tomtom, comes the thrilling rhythm hour after hour, of a double row of paddles tearing and eddying through water in unison, not only the thump and splash from the dugouts of tropical savages but the deep-dipped rush and swirl from bark canoes. This is the obvious, the much-described, but how many of us have listened for, and heard, the low, sibilant swish of the blades through the air, as they reach forward for the next stroke. Until mind and ear are focussed it is inaudible, but when once caught it outsings more blatant sounds of water and voice. The blind spots of our perceptions conceal many phases of delicate beauty in the things around us, aspects which are dulled by the opacity of familiarity, passed over by the unseeing activity of our surface-skimming minds.

The living leaf—both singly and in foliage mass—has been epitaphed, eulogized, sung, praised and similed for centuries, but except for occasional references to the "sere and yellow leaf," dying, falling and dead leaves have been left where they lie, with only the incense of their funeral pyres woven into the haze of Indian Summer.

I have seen an orang-utan build him a sleeping platform of leaves in less than three minutes, so it is not improbable that the first artificial home our more direct ancestors knew was a leafy nest. Leaves at least formed the sole clothing of our early parents, according to Scripture, and from nursery days we have always known that falling leaves were a shroud for the babes in the wood. More than this, botanists tell us that the leaf is the foundation of flower and fruit, so that it was really only a mass of highly specialized leaves which introduced Newton to gravitation.

But the importance and interest of falling leaves in this world needs no brief from me. I merely want to know them better for my own pleasure, I wish to hear and see and feel them, and so I leave my laboratory after a day of intensive technical work and slip into the jungle, where millions of leaves are falling during my lifetime, and hundreds of millions fell before I was born.

I am sitting at the edge of a tropical swamp and for the moment trying to close my mind and sense

FALLING LEAVES

to the sounds and sights of birds and insects, and focus on leaves, and especially dead ones. This is no more difficult than it would have been to have forgotten Caruso and the orchestra in order to meditate on the kind of wood of which the chairs were fashioned.

Further than this I am putting out of my mind the letters L E A V E S and thinking of them innominately as a vast multitude of spread-out sheets of green and brown tissue. They are really the jungle, for without them it would be like the bare masts and rigging of a vessel. High overhead beyond the clouds of chlorophyll are other white clouds of moisture, driven swiftly westward by the steady trade-wind. Around me the air is as quiet as in a room, and, as so often the case, of just the right temperature to be forgotten, neither too hot nor too cold, a distinct effort being necessary to realize that I am not in some great enclosed chamber; so calm and equable are the surroundings.

It is the dry season, and the short daily shower does little to soften the crackle of the fallen leaves. Even after a month of heavy unseasonable rain, when our records show that it is the dry season, the noise of treading on the jungle floor reveals the actual lack of humidity at times other than actual precipitation. Now and then, near my feet, a leaf

JUNGLE DAYS

draws its edges together, turns a little and rustles gently all by itself as if even in death it dreamed of some pleasant trifle, something which would please a green leaf, in sunlight, swaying high in air. Then, like a crumpled bit of paper in a wastebasket, it settles lower among its fallen fellows. Here it will wait patiently for the impact of the heavy rains, three or four months hence, to soften its stiff, crinkling tissues, and re-mold it into incarnations of other leaves to come.

Fallen leaves have a wind song all their own which is to be heard only when listened for consciously. When a fitful breeze is blowing, if the ear is held close to the ground, a low intermittent clatter and shuffling is audible, with occasionally a real rustle as a delicately balanced leaf is blown over. Stand up and the carpet of dead leaves becomes silent, their gentle talk lost in the hubbub of living, moving foliage.

In this quiet, cool swamp I am impressed with the vast numbers of leaves which have started to fall but have not reached the earth. Some have landed in crotches, or become entangled in masses of vines, others have driven their stems clear through the live tissue of leaves in their downward path and hang dangling. Just above me a living and a dead palmated frond have their leafy fingers

FALLING LEAVES

intertwined like the outer points of fighting buckles, with no chance of release until the death and fall of the second leaf.

As I watched, three leaves fell, each with characteristic motion. I once made a key to more than a dozen kinds of jungle trees, based on the way the leaves fell, and to anyone who wishes to enter an untrodden botanical field I commend this idea. The third leaf fluttered and eddied, fighting with all its expanse of plane against the pull of gravitation, and at the very last, came to rest on a mattress of fern frond—a respite merely, for the first real gust would send it to the ground. As it touched the fern a butterfly rose, a black heliconian, with a large red spot on each wing. Its flight was astonishingly like that of the descending leaf, a tremulous fluttering just carrying it along, now rising, now descending—a flight wholly deceiving, for these butterflies can thread the mazes of jungle vines all day without tiring. But this butterfly was also like the leaf in its sear and faded garb. The wings were frayed and torn—the black was a thread-bare brown, the red weathered to faded salmon, and the seams of its wings showed plainly. Life was nearly over, yet weak as it was, it would probably die no violent death. The most awkward bird or predatory insect could catch it at will, yet it flew slowly along, unmolested by jacamars and cuckoos, dragon and robber flies. Its conspicuous colors and slow, tantalizing flight, like all else in the jungle, had a reason—it was its own advertisement of inedibility. Soon, however, this Wandering Jew of a butterfly would slip from its sleeping porch, and, like the fluttering leaf, make a last ineffectual struggle against the pull of earth and its wings would lie among the leaves.

Before the butterfly passed from view, I was startled by a sudden, rough rip of sound,—and just overhead a macaw put all the harshness of its beak and the blatancy of its coloring into its voice, and almost the leaves around me seemed to rustle. Into a clear space of sky four great, flame-winged birds passed, and with flight direct as arrows, but otherwise exactly like the falling leaf and the butterfly, they vibrated northward.

Without intention, but very happily, I found I had chosen my seat between extremes in leaves. Close along one side lay a fallen leaf which began eight feet behind and extended twenty-three feet in front,—thirty-one feet of palm frond. In its fall it had crushed several young mora saplings and many lesser growths. The least movement near it aroused a crashing which could be heard to the river. The leaflets, two hundred in number, lay

FALLING LEAVES

stretched out four to six feet on each side, and the mighty stem was like a length of channel iron, with edges sharp as razors. It was parched and shrunken and had probably hung dead for a long time before it fell. A billion ordinary leaves fall unnoticed in the tropics, while in the North we lump this vast assemblage of happenings under the one word "autumn." But the fall of a palm leaf is an event. Once as I was leaving my Station for a trip north, I noticed that one of the leaves of our sentinel cuyuru palm was drooping and browned. Months later when I returned, it was still hanging, and two weeks afterwards fell in the night with a crash which wakened us all. Dynasties of history might be dated by the falling of such a leaf, and if I could have been present at the dropping of all the leaves of my palm, whose scars were still so plain, there would be material for an epic. The remark of Charles the Second on his deathbed could be applied to the dead leaf at my side, for these gigantic fronds grow and live their lives much more rapidly than they die and disintegrate. Years from now I could probably find traces of the reinforced cellulose-hardened main stem.

And now my faded and forlorn heliconian butterfly fluttered again toward me, and almost alighted on this paper, but turning at the last moment, it rose a bit, and came to rest at my elbow, on a stem lined with small leaflets. Hardly had the insect furled its wings, when it fluttered and took to flight again. The cause delighted me beyond measure,— it had been unseated and frightened by the movement of a living leaf! At the impact of its delicate feet, the leaflets of the sensitive plant closed abruptly together and the stem sank. So exquisite was the reaction that the several leaflets beyond the insect were unmoved. A few seconds later while I was still watching, an adjoining twiglet closed every one of its leaflets and dropped 120° upon its parent branch. Nothing had touched it, no breath of air had moved it. I was puzzled. Lifting it very gently, it broke off and fell to the ground, green, fresh,—as far as I could see quite without cause. I picked it up and examined the base and there I found the source of the trouble. A tiny beetle had cut it almost off, and the slight fall of the twig, together with my touch had parted the few remaining fibres. The beetle was very small and must have been laboring for a long time, and it was a mystery why the featherdom tread of a butterfly's feet had accomplished what the hacking and sawing of the beetle's jaws had not.

All the leaves on the mimosa would not have equalled one of the lesser leaflets of the palm frond, and on the ground they were almost invisible, sinking almost at once into the mold. The sensitive leaves had the semblance of animal nerves and movement; the palm leaf would have brained me if it had fallen while I passed beneath.

In these jungles a falling leaf has a whole scale of sounds, as it runs the descending gamut of collisions. From the top of a tall tree a leaf may take fifteen or twenty seconds to reach the earth, disregarding the very good chance of lodgment, and each touch of vine, leaves,—living and dead,—the caroming off of branches and ripping through thorns, gives forth a different sound, of which our poor ears can distinguish very few, and which our language, spoken and written, is wholly helpless in reproducing. I would like very much to find a word or sound which would bring to mind the fall of a leaf upon leaves. I know it perfectly—the generic timbre—the composite echo etched into my mind by a thousand conscious listenings. But it will not get past my consciousness to my lips, and utterly refuses to siphon down my arm and pen.

Fallen leaves are of tremendous importance to those of us who do much hunting in the jungle, and chiefly on account of their susceptibility to moisture in the air. In the wet season it is possible to creep up to some of the wariest of animals, the thick mat of soft, damp leaves forming an admirable muffler. In the dry season this is hopeless, every step is ascream with crackling, and only when a leaf-rattling breeze is blowing can one pass through the jungle without blatant advertisement. This, however, is of slight assistance in hunting, for the blowing of the leaves conceals as well the audible whereabouts of the game. When the fallen leaves are dry the only method is to walk to some favorable spot, and there sit and wait for approaching or passing animals to register their footfalls. In estimating the abundance of jungle life I have constantly to check a tendency to underestimate numbers in the wet season. Ameiva lizards appear to be many times as abundant in times of drought, crashing along with the noise of a peccary, yet they have no season of æstivation, but only of silent progress.

We do not realize the acuteness of hearing of wild animals until we try to stalk them over dry leaves. A giant leaf may crash down from branch to branch and never cause a curassow or deer to start. I have seen a labba feeding in late afternoon under a nut tree when a whole branch with clusters of dead leaves hurtled to earth a few yards away, and the big, spotted rodent merely glanced up,

FALLING LEAVES

casually munching as it looked. My next step slipped an inch sideways and crumbled a tiny leaf crust, and without a second's investigation the animal gave one terrified squeal and fled headlong.

There are silent and there are boisterous leaves. Some, with finely pinnated foliage, have a pact of silence with the elements, from which wind and rain strive in vain to awaken them. Even when these filigree ones are dead and cling long to the branches, they give before the blasts, they let the rain drip from their finger tips without a sound. But a single, half-loose cecropia frond can imitate a rainstorm, the roar of a flushed covey of pheasants or a passing troop of monkeys, all by itself. More than this, it will begin uncannily to quiver and shake and rattle wildly about, while every adjacent leaf dangles as silently as if painted. Thus does its sensitive balance and crinkled shard betray the wandering little wind spouts which are born deep in the jungle, and, like other aquatic cousins, stretch straight upward in a tiny, clean-cut whorl of air.

A book could be written upon burning leaves—how they meet their cremation, how they curl when this new, devastating long-bottled-up sun heat chars their tissues. How they shout and crack in the wind of their own swan song, and how they look when the heat and roar have passed and the cold ash remains. A month of drought at Kartabo once made the thick mat of bamboo leaves about the compound considerable of a menace. So we had a great raking and bonfire of the ten million and one elongated slivers of pale brown leaves. (Even the color of dead leaves, like the plumage of hen pheasants, is far more subtle and beautiful than we suspect, for after the above sentence, I try to match a dead bamboo leaf color in Ridgway's color book and fail utterly. It lies between vinaceous-buff and olive-buff and is of no human-named color.)

The ashy souls of leaves differ to as great a degree as do their shapes and life-greens. Some are so ethereal that they vanish in a curl of faint blue smoke and leave scarcely a trace of ponderable greyness. The bamboos are far otherwise. There is nothing quiet or sad about their cremation. They snap and crackle joyously in the flames, with more gust than ever they rattled in the trade winds. And indeed their passing is far less of a radical change than for most leaves. They are so surcharged with silica that the alchemy of glowing heat merely alters their hue to silvery white, and when the furnace of their tissues has cooled, they lie unchanged in shape and outline. A heavy rain or big wind shatters this crystalline ghost of a *feuille,* and the various salts

"The jungle *du printemps eternel*"

"In the sunshine and warmth of the mangrove tangle"

FALLING LEAVES

are washed into the soil, ready for their next great adventure.

Before I lived under bamboos I never realized how friendly fallen leaves could be. Trees with heavy, leaded-stemmed leaves drop them straight to the ground. But bamboo leaves are like zeppelins when they are launched and, with the slightest breeze float along on even keels, drifting sometimes far into the laboratory. When at tea one day I idly watched a leaf dangling high up from one of the lofty stems, so far away I could not tell whether it was brown or green. A slight gust came and it broke off and, revolving slowly, scaled obliquely down, through the verandah and launched in my teacup.

These leaves register very accurately the force of the wind, and I have seen a thick bed of ashes of burned bamboo leaves studded thickly as a porcupine's skin with the javelins of recent falls, two lots having speared the ashes at different angles. One was almost upright, having landed in a gentle wind that afternoon, the other at an oblique angle, after volplaning on the stronger trades of morning.

Leaves in death still mirror many of the characteristics of their living fellows. In the tropics a host of plants flower once or at most twice a year, but attract insects at all times by setting forth a

JUNGLE DAYS

little bowl of nectar on each leaf stalk. I have observed a small bush with forty-nine leaves and counted nine and forty ants thereon, one guest to each nectar-cup,—each having visited, sipped and remained—perhaps by their jealous gormandizing keeping away other more harmful insects. On fallen leaves the sides of the bowls still seem to contain some sweetness, and to these come other ants (as we used to love to scrape the emptied ice-cream freezer), who gnaw eagerly at the shrivelled cups and the sweet crusts which have fallen from the table of the jungle.

There are parts of jungle clearings which I hardly know in early morning, while their foliage is still asleep. Some leaves are surprisingly drowsy and not until the sun actually fillips them with its beams do they raise their heads, twist on their stalks in a leafy yawn, and eat to their daily stint in their chlorophyll factory. These leaves die in the position of sleep, so that if we had a fallen twigful we would know their somnolent attitude in life.

By far the phase of dominant interest in fallen or dead leaves is the part they have played in the evolution of animal life. If we can infer the position of sleep from that in death, how vastly greater is possible the reconstruction of dead vegetation from living creatures of the jungle. If every leaf

FALLING LEAVES

and twig, flower and fruit, branch and trunk were to vanish suddenly from the earth, their memory would remain deeply impressed in form, size, movement, pattern and color of a host of creatures, while we would still have even the jungle lights and shadows etched upon fur and feathers. As we go down the scale in life we find more and more marvels of resemblance, and it would be an easy matter to reconstruct an entire plant of animals. I have caught monster walking-stick insects over a foot in length, which were dead wood to the keenest eye. Smaller ones carry the resemblance to an inordinate extreme. Not only do they look like twigs and stems, but they *act* like them, clinging with four feet and dangling the other two out in midair, while every now and then the whole insect sways gently, as does a tiny twig moved by a breath. Things such as this make a scientist's work wonderful and holy beyond Bryan's utmost conception of these words.

From day to day in the jungle I add to my animal-plants. I discover giant katy-dids so green and flat, so veined and stemmed, that no passing observer could say, "This is leaf, this insect." Others have spoiled the symmetry and perfection of their sham chlorophyll with simulated holes, and apparent tears and spots of fungi, and the droppings of birds. All the diseases, parasites and injuries of leaves have been photographed upon the wings of insects, in unconscious endeavor to escape observation. At this point we come upon interactions, complications, subtleties of great delicacy, such as are shown by mantids, or "rar'hor'ses" as they are called in the Southern States. These are incarnated, material sophists, camouflaged under chlorophyll color not for protection but for attack. As the white fox creeps upon the white ptarmigan over the white snow, so here in the tropics, the mantids re-enact a similar, but viridescent drama.

Passing on from growing leaves we find flower bugs and orchid spiders, the latter being forced to conceal their brilliant pigments in the shadow of under-leaf, until some particular blossom appears. Then, with their colors and patterns so exact that they might have been fashioned in the same petal shop the spiders take their place on or near the flowers. Some even eat away the heart of the blossom, substituting their stamen leg and pistil palpi, and with the unharmed nectary still giving forth perfume, these deadly frauds of flowers await the visiting bees.

Caterpillars gnaw out bits of leaf and then fill up the space with their own painted bodies, but butterflies and moths are the veritable reflections

FALLING LEAVES

of leaves, they would indeed be naked and blatant to the world were foliage to vanish. Here again not only are color and pattern invoked but even the movement in falling. I have had a brown butterfly flutter in short, oblique eddies to my feet, and there alight warelessly and sway from side to side. Dozens of times I have crept up and enmeshed a dead leaf in my net, and as many times have brushed heedlessly by a dead leaf only to have it take wings to itself and fly away.

Two adventures which befell me yesterday had to do with leaves, and touched the extremes of the gamut of an explorer's life—from the danger of death to the glory of new discovery. Every morning a bird had been calling from a certain tree-top —a short, raucous, unpleasant call, but a new one. So ventriloquial was it that it had wholly baffled me. Only by triangulation, the successive focussing from three distant points, could I ever hope to find it. I was creeping slowly on my second lap, lifting my feet high to clear twigs and vines, when something drew my eyes from the tree overhead to the dead leaves below. This has happened to me perhaps a score of times and I hope will continue in the future—the sudden, inexplicable perception of a poisonous snake on which my foot is about to descend. A large fer-de-lance, more like dead leaves

JUNGLE DAYS

than the leaves themselves, was coiled less than two feet away. On its scales it mirrored the brown dead leaves, the dark fungus spots, the shadows of the curled-up edges, the high lights of the burnished surface sheen. Optically there was no interruption of the floor of dead foliage; actually a horrible death lay twelve inches beneath my upraised foot. The lethal mat was coiled as evenly as a rope on a battleship and in the exact center lay the arrow head with its unwinking eyes and the flickering tongue. As I withdrew my foot and began to breathe again, I forgot my raucous-voiced bird and sat down to ponder this. I took my strong butterfly net and drew the netting taut across the ring and behind this barrier I slowly approached. Closer and closer I drew until I could see the slit-like pupil and the green and livid mottling of the iris. When I almost touched the sharp snout with the other side of the mesh, I sniffed carefully and repeatedly, dulling every other sense but that of smell. There came to my nostrils a faint but distinct odor, an unpleasant musk, which, once detected, remained vivid. It was a faint adumbration of that strong, repulsive smell which permeates the cage where one of these reptiles is confined, and I believe that, without invoking any more radically psychic process, my attention is attracted and focussed at these times by the faint, unconsciously stimulating odor of the snake on the jungle floor. I cannot otherwise explain my invariable detection at the last minute, of creatures who more than any others are of the leaves, leafy.

My second adventure was also a thrilling one but from a wholly different point of view. I was walking along a trail after a shower, looking idly at a big, palmated leaf at my very elbow when there suddenly materialized upon it a large lizard. It was one of the most beautiful of all lizards and fortunately had been named with imagination—*Polychrus marmoratus*—the many-colored Marble One. It was sprawled flat upon the great green expanse, its scales shimmering leaf-green with enough spots here and there to be a convincing portion of the full-grown, insect-defaced foliage. I leaned toward it and it began slowly to creep away. The long, slender tail was curled and twisted into a lifeless tendril, and the toes dangled half in midair like no imaginable piece of any live reptile. Progress was by means of the forefeet alone, one after the other being pushed ahead stealthily, taking hold and dragging the rest of the creature onward. The body, hind legs and tail simply scraped over the leaf.

When it reached the thick, brown twig, magic

JUNGLE DAYS

began before our eyes—for fortunately I had two companions to share this wonder. As it left the green tissue and crawled slowly out along the twig its course was traceable not only by its position in space, but by most exquisitely adjusted and timed pigmental change,—at the exact edge of the leaf the green gradually faded and a wave of brown swept down the reptile. Never have I seen a more perfect use of obliterative color. In captivity these polychrus will often run through their whole little pigmental gamut from mere emotion, or light and shadow. The whole soul of my lizard on the leaf was concentrated in his half-closed eyes watching my every motion, yet it must have been through the eye alone that the amazingly accurate somatic color change was dictated and regulated. Here was surely the ultimate example of vegetable imitation, twigs, leaves—both green and brown—tendril swaying movement, all in one organism. Not for anything would I have betrayed the lizard's trust in the magnificent shield which nature had built up about it. We pretended to be completely deceived and left it—an irregular bit of half-greenness on the second leaf, and half brownness on the twig.

A classic volume will some day be written on the adventures of fallen leaves, for when a leaf has evaded the inroads of insects and fungi, has resisted wind and rain, succumbing finally to the pull of gravitation, there awaits it, in addition to ultimate mold and desiccation, a host of possible adventures on the jungle floor.

With all my desire to clothe the fallen leaf with dramatic interest and an abstract vitality, my first and last thoughts are those of sadness. Alien as I am to these tropical jungles, a mere transient injection from the North, the sear and yellow leaf means to me the end of a season, of a year—a very appreciable fraction of lifetime—and even in this evergreen land, this jungle *du printemps éternel*, the dead leaf eddying to earth is a sad and a tragic happening.

V

THE JUNGLE SLUGGARD

SLOTHS have no right to be living on the earth today; they would be fitting inhabitants of Mars, where a year is over six hundred days long. In fact they would exist more appropriately on a still more distant planet where time—as we know it —creeps and crawls instead of flies from dawn to dusk. Years ago I wrote that sloths reminded me of nothing so much as the wonderful Rath Brother athletes or of a slowed-up moving picture, and I can still think of no better similes.

Sloths live altogether in trees, but so do monkeys, and the chief difference between them would seem to be that the latter spend their time pushing against gravitation while the sloths pull against it. Botanically the two groups of animals are comparable to the flower which holds its head up to the sun, swaying on its long stem, and, on the other hand, the over-ripe fruit dangling heavily from its

THE JUNGLE SLUGGARD

base. We ourselves are physically far removed from sloths—for while we can point with pride to the daily achievement of those ambulatory athletes, floor-walkers and policemen, yet no human being can cling with his hands to a branch for more than a comparatively short time.

Like a rainbow before breakfast, a sloth is a surprise, an unexpected fellow breather of the air of our planet. No one could prophesy a sloth. If you have an imaginative friend who has never seen a sloth and ask him to describe what he thinks it ought to be like, his uncontrolled phrases will fall far short of reality. If there were no sloths, Dunsany would hesitate to put such a creature in the forests of Mluna, Marco Polo would deny having seen one, and Munchausen would whistle as he listened to a friend's description.

A scientist—even a taxonomist himself—falters when he mentions the group to which a sloth belongs. A taxonomist is the most terribly accurate person in the world, dealing with unvarying facts, and his names and descriptions of animals defy discretion, murder imagination. Nevertheless when next you see a taxonomist disengaged, approach him boldly and ask him in a tone of quarrelsome interest to what order of Mammalia sloths belong. If an honest conservative he will say,

"Edentata," which, as any ancient Greek will tell you, means a toothless one. Then if you wish to enrage and nonplus the taxonomist, which I think no one should, as I am one myself, then ask him Why? or, if he has ever been bitten by any of the eighteen teeth of a sloth?

The great savant Buffon in spite of all his genius, fell into most grievous error in his estimation of a sloth. He says, "The inertia of this animal is not so much due to laziness as to wretchedness; it is the consequence of its faulty structure. Inactivity, stupidity, and even habitual suffering result from its strange and ill-constructed conformation. Having no weapons for attack or defense, no mode of refuge even by burrowing, its only safety is in flight. . . . Everything about it shows its wretchedness and proclaims it to be one of those defective monsters, those imperfect sketches, which Nature has sometimes formed, and which, having scarcely the faculty of existence, could only continue for a short time and have since been removed from the catalogue of living beings. They are the last possible term amongst creatures of flesh and blood, and any further defect would have made their existence impossible."

If we imagine the dignified French savant himself, naked, and dangling from a lofty jungle

THE JUNGLE SLUGGARD

branch in the full heat of the tropic sun, without water and with the prospect of nothing but coarse leaves for breakfast, dinner and all future meals, an impartial onlooker who was ignorant of man's normal haunts and life could very truthfully apply to the unhappy scientist, Buffon's own comments. All of his terms of opprobrium would come home to roost with him.

A bridge out of place would be an absolutely inexplicable thing, as would a sloth in Paris, or a Buffon in the trees. As a matter of fact it was only when I became a temporary cripple myself that I began to appreciate the astonishing lives which sloths lead. With one of my feet injured and out of commission I found an abundance of time in six weeks to study the individuals which we caught in the jungle near by. Not until we invent a superlative of which the word "deliberate" is the positive can we define a sloth with sufficient adequateness and briefness. I dimly remember certain volumes by an authoress whose style pictured the hero walking from the door to the front gate, placing first the right, then the left foot before him as he went. With such detail and speed of action might one write the biography of a sloth.

Ever since man has ventured into this wilderness, sloths have aroused astonishment and comment.

JUNGLE DAYS

Four hundred years ago Gonzala de Oviedo sat him down and penned a most delectable account of these creatures. He says, in part: "There is another strange beast the Spaniards call the Light Dogge, which is one of the slowest beasts and so heavie and dull in mooving that it can scarsely goe fiftie pases in a whole day. Their neckes are high and streight, and all equall like the pestle of a mortar, without making any proportion of similitude of a head, or any difference except in the noddle, and in the tops of their neckes. They have little mouthes, and moove their neckes from one side to another, as though they were astonished: their chiefe desire and delight is to cleave and sticke fast unto Trees, whereunto cleaving fast, they mount up little by little, staying themselves by their long claws. Their voice is much differing from other beasts, for they sing only in the night, and that continually from time to time, singing ever six notes one higher than another. Sometimes the Christian men find these beasts, and bring them home to their houses, where also they creepe all about with their natural slownesse. I could never perceive other but that they love onely of Aire: because they ever turne their heads and mouthes toward that part where the wind bloweth most, whereby may be considered that they take most pleasure in the Aire. They bite not,

THE JUNGLE SLUGGARD

nor yet can bite, having very little mouthes: they are not venemous or noyous any way, but altogether brutish, and utterly unprofitable and without commoditie yet known to men."

It is difficult to find adequate comparisons for a topsy-turvy creature like a sloth, but if I had already had synthetic experience with a Golem, I would take for a formula the general appearance of an English sheep dog, giving it a face with barely distinguishable features and no expression, an inexhaustible appetite for a single kind of coarse leaf, a gamut of emotions well below the animal kingdom, and an enthusiasm for life excelled by a healthy sunflower. Suspend this from a jungle limb by a dozen strong hooks, and—you would still have to see a live sloth to appreciate its appearance.

At rest, curled up into an arboreal ball, a sloth is indistinguishable from a cluster of leaves; in action, the second hand of a watch often covers more distance. At first sight of the shapeless ball of hay, moving with hopeless inadequacy, astonishment shifts to pity, then to impatience and finally, as we sense a life of years spent thus, we feel almost disgust. At which moment the sloth reaches blindly in our direction, thinking us a barren, leafless, but perhaps climbable tree, and our emotions change again, this time to sheer delight as a tiny

JUNGLE DAYS

infant sloth raises its indescribably funny face from its mother's breast and sends forth the single tone, the high, whistling squeak, which in sloth intercourse is song, shout, converse, whisper, argument and chant. Separating him from his mother is like plucking a bur from one's hair, but when freed, he contentedly hooks his small self to our clothing and creeps slowly about.

Instead of reviewing all the observations and experiments which I perpetrated upon sloths, I will touch at once the heart of their mysterious psychology, giving in a few words a conception of their strange, uncanny minds. A bird will give up its life in defending its young; an alligator will not often desert its nest in the face of danger; a male stickleback fish will intrepidly face any intruder that threatens its eggs. In fact, at the time when the young of all animals are at the age of helplessness, the senses of the parents are doubly keen, their activities and weapons are at greatest efficiency for the guarding of the young and the consequent certainty of the continuance of their race.

The resistance made by a mother sloth to the abstraction of its offspring is chiefly the mechanical tangling of the young animal's tiny claws in the long maternal fur. I have taken away a young sloth and hooked it to a branch five feet away. Be-

THE JUNGLE SLUGGARD

ing hungry it began at once to utter its high, penetrating penny whistle. To no other sound, high or low, with even a half tone's difference does the sloth pay any heed, but its dim hearing is attuned to just this vibration. Slowly the mother starts off in what she thinks is the direction of the sound. It is the moment of moments in the life of the young animal. Yet I have seen her again and again on different occasions pass within two feet of the little chap, and never look to right or left, but keep straight on, stolidly and unvaryingly to the high jungle, while her baby, a few inches out of her path, called in vain. No kidnapped child hidden in mountain fastness or urban underworld was ever more completely lost to its parent than this infant, in full view and separated by only a sloth's length of space.

A gun fired close to the ear of a sloth will usually arouse not the slightest tremor; no scent of flower or acid or carrion causes any reaction; a sleeping sloth may be shaken violently without awakening, the waving of a scarlet rag, or a climbing serpent a few feet away brings no gleam of curiosity or fear to the dull eyes; an astonishingly long immersion in water produces discomfort but not death. When we think what a constant struggle life is to most creatures, even when they are equipped with the keenest of senses and powerful means of offense, it seems incredible that a sloth can hold its own in this overcrowded tropical jungle.

From birth to death it climbs slowly about the great trees, leisurely feeding, languidly loving, and almost mechanically caring for its young. On the ground a host of enemies await it, but among the higher branches it fears chiefly occasional great boas, climbing jaguars and, worst of all, the mighty talons of harpy eagles. Its means of offense is a joke—a slow, ineffective reaching forward with open jaws, a lethargic stroke of arm and claws which anything but another sloth can avoid. Yet the race of sloths persists and thrives, and in past years I have had as many as eighteen under observation at one time.

A sloth makes no nest or shelter; it even disdains the protection of dense foliage. But for all its apparent helplessness it has a *cheval-de-frise* of protection which many animals far above it in intelligence might well envy. Its outer line of defense is invisibility—and there is none better, for until you have seen your intended prey you can neither attack nor devour him. No hedgehog or armadillo ever rolled a more perfect ball of itself than does a sloth, sitting in a lofty, swaying crotch with head and feet and legs all gathered close to-

THE JUNGLE SLUGGARD

gether inside. This posture, to an onlooker, destroys all thought of a living animal, but presents a very satisfactory white ants' nest or bunch of dead leaves. If we look at the hair of a sloth we will see small, grey patches along the length of the hairs—at first sight bits of bark and débris of wood. But these minute, scattered particles are of the utmost aid to this invisibility. They are a peculiar species of alga or lichen-like growth which is found only in this peculiar haunt, and when the rains begin and all the jungle turns a deep, glowing emerald, these tiny plants also react to the welcome moisture and become verdant—thus throwing over the sloth a protecting, misty veil of green.

Even we dull-sensed humans require neither sight nor hearing to detect the presence of an animal like the skunk; in the absolute quiet and blackness of midnight we can tell when a porcupine has crossed our path, or when there are mice in the bureau drawers. But a dozen sloths may be hanging to the trees near at hand and never the slightest whiff of odor comes from them. A baby sloth has not even a baby smell, and all this is part of the cloak of invisibility. The voice, raised so very seldom, is so ventriloquial, and possesses such a strange, unanimal-like quality that it can never be a guide to the location, much less to the identity of the author. Here we have three senses, sight, hearing, smell, all operating at a distance, two of them by vibrations, and all leagued together to shelter the sloth from attack.

But in spite of this dramatic guard of invisibility the keen eyes of an eagle, the lapping tongue of a giant boa, and the amazing delicacy of a jaguar's sense of smell break through at times. The jaguar scents sign under the tree of the sloth, climbs eagerly as far as he dares and finds ready to his paw the ball of animal unconsciousness; a harpy eagle half a mile above the jungle sees a bunch of leaves reach out a sleepy arm and scratch itself—something clumps of leaves should not do. Down spirals the great bird, slowly, majestically, knowing there is no need of haste, and alights close by the mammalian sphere. Still the sloth does not move, apparently waiting for what fate may bring —waiting with that patience and resignation which comes only to those of our fellow creatures who cannot say, "I am I!" It seems as if Nature had deserted her jungle changeling, stripped now of its protecting cloak.

The sloth however has never been given credit for its powers of passive resistance, and now, with its enemy within striking distance, its death or even injury is far from a certainty. The crotch which

THE JUNGLE SLUGGARD

the sloth chooses for its favorite outdoor sport, sleep, is unusually high up or far out among the lesser branches, where the eight claws of the eagle or the eighteen of a jaguar find but precarious hold. In order to strike at the quiescent animal the bird has to relinquish half of its foothold, the cat nearly one quarter. If the victim were a feathery bush turkey or a soft-bodied squirrel, one stroke would be sufficient, but this strange creature is something far different. In the first place it is only to be plucked from its perch by the exertion of enormous strength. No man can seize a sloth by the long hair of the back and pull it off. So strong are its muscles, so vise-like the grip of its dozen talons that either the crotch must be cut or broken off or the long claws unfastened one by one. Neither of these alternatives is possible to the attacking cat or eagle. They must depend upon crushing or penetrating power of stroke or grasp.

Here is where the sloth's second line of defense becomes operative. First, as I have mentioned, the swaying branch and dizzy height is in his favor, as well as his immovable grip. To begin with the innermost defenses, while his jungle fellows, the ring-tailed and red howling monkeys, have thirteen ribs, the sloth may have as many as twenty; in the latter animal they are, in addition, unusually broad

JUNGLE DAYS

and flat, slats rather than rods. Next comes the skin which is so thick and tough that many an Indian's arrow falls back without even scratching the hide. The skin of the unborn sloth is as tough and strong as that of a full-grown monkey. Finally we have the fur—two distinct coats, the under one fine, short and matted, the outer long, harsh and coarse. Is it any wonder that, teetering on a swaying branch, many a jaguar has had to give up after frantic attempts to strike his claws through the felted hair, the tough skin and the bony lattice-work which protect the vitals of this Edentate bur!

Having rescued our sloth from his most immediate peril let us watch him solve some of the very few problems which life presents to him. Although the cecropia tree, on the leaves of which he feeds, is scattered far and wide through the jungle, yet sloths are found almost exclusively along river banks, and, most amazingly, they not infrequently take to the water. I have caught a dozen sloths swimming rivers a mile or more in width. Judging from the speed of short distances, a sloth can swim a mile in three hours and twenty minutes. Their thick skin and fur must be a protection against crocodiles, electric eels and perai fish as well as jaguars. Why they should ever wish to swim

THE JUNGLE SLUGGARD

across these wide expanses of water is as inexplicaable as the migration of butterflies. One side of the river has as many comfortable crotches, as many millions of cecropia leaves and as many eligible lady sloths as the other! In this unreasonable desire for anything which is out of reach sloths come very close to a characteristic of human beings.

Even in the jungle sloths are not always the static creatures which their vegetable-like life would lead us to believe, as I was able to prove many years ago. A young male was brought in by Indians and after keeping it a few days I shaved off two patches of hair from the center of the back. and labelling it with a metal tag I turned it loose. Forty-eight days later it was captured near a small settlement of bovianders several miles farther up and across the river. During this time it must have traversed four miles of jungle and one of river.

The principal difference between the male and female three-toed sloths is the presence on the back of the male of a large, oval spot of orange-colored fur. To any creature of more active mentality such a minor distinction must often be embarrassing. In an approaching sloth, walking upside-down as usual, this mark is quite invisible, and hence every meeting of two sloths must contain much of delightful uncertainty, of ignorance whether the encounter presages courtship or merely gossip. But color or markings have no meaning in the dull eyes of these animals. Until they have sniffed and almost touched noses they show no recognition or reaction whatever.

I once invented a sloth island—a large circle of ground surrounded by a deep ditch, where sloths climbed about some saplings and ate, but principally slept, and lived for months at a time. This was within sight of my laboratory table, so I could watch what was taking place by merely raising my head. Some of the occurrences were almost too strange for creatures of this earth. I watched two courtships, each resulting in nothing more serious than my own amusement. A female was asleep in a low crotch, curled up into a perfect ball deep within which was ensconced a month-old baby. Two yards overhead was a male who had slept for nine hours without interruption. Moved by what, to a sloth, must have been a burst of uncontrollable emotion, he slowly unwound himself and clambered downward. When close to the sleeping beauty he reached out a claw and tentatively touched a shoulder. Even more deliberately she excavated her head and long neck and peered in every direction but the right one. At last she perceived her suitor

THE JUNGLE SLUGGARD

and looked away as if the sight was too much for her. Again he touched her post-like neck, and now there arose all the flaming fury of a mother at the flirtatious advances of this stranger. With incredible slowness and effort she freed an arm, deliberately drew it back and then began a slow forward stroke with arm and claws. Meanwhile her gentleman friend had changed his position so the blow swept, or, more correctly passed, through empty air, the lack of impact almost throwing her out of the crotch. The disdained one left with slowness and dignity—or had he already forgotten why he had descended?—and returned to his perch and slumber, where I am sure, not even such active things as dreams came to disturb his peace.

The second courtship advanced to the stage where the Gallant actually got his claws tangled in the lady's back hair before she awoke. When she grasped the situation she left at once and clambered to the highest branch tip followed by the male. Then she turned and climbed down and across her annoyer, leaving him stranded on the lofty branch looking eagerly about and reaching out hopefully toward a big, green iguana asleep on the next limb in mistake for his fair companion. For an hour he wandered languidly after her, then gave it up and went to sleep. Throughout these

JUNGLE DAYS

and other emotional crises no sound is ever uttered, no feature altered from its stolid repose. The head moves mechanically and the dull eyes blink slowly, as if striving to pierce the opaque veil which ever hangs between the brain of a sloth and the sights, sounds and odors of this tropical world. If the orange back spot was ever of any use in courtship, in arousing any emotion æsthetic or otherwise, it must have been in ages long past when the ancestors of sloths, contemporaries of their gigantic relatives the Mylodons, had better eyesight for escaping from sabre-toothed tigers, than there is need today.

The climax of a sloth's emotion has nothing to do with the opposite sex or with the young, but is exhibited when two females are confined in a cage together. The result is wholly unexpected. After sniffing at one another for a moment, they engage in a slowed-up moving-picture battle. Before any harm is done one or the other gives utterance to the usual piercing whistle and surrenders. She lies flat on the cage floor and offers no defense while the second female proceeds to claw her, now and then attempting, usually vainly, to bite. It is so unpleasant that I have always separated them at this stage, but there is no doubt that in every case the unnatural affray would go on until the victim was

THE JUNGLE SLUGGARD

killed. In fact I have heard of several instances where this actually took place.

A far pleasanter sight is the young sloth, one of the most adorable balls of fuzzy fur imaginable. While the sense of play is all but lacking his trustfulness and helplessness are most infantile. Every person who takes him up is an accepted substitute for his mother and he will clamber slowly about one's clothing for hours in supreme contentment. One thing I can never explain is that on the ground the baby is even more helpless than his parents. While they can hitch themselves along, body dragging, limbs outspread, until they reach the nearest tree, a young sloth is wholly without power to move. Placed on a flat bit of ground it rolls and tumbles about, occasionally greatly encouraged by seizing hold of its own foot or leg under the impression that at last it has encountered a branch.

Sloths sleep about twice as much as other mammals and a baby sloth often gets tired of being confined in the heart of its mother's sleeping sphere, and creeping out under her arm will go on an exploring expedition around and around her. When over two weeks old it has strength to rise on its hind legs and sway back and forth like nothing else in the world. Its eyes are only a little keener than those of the parent and it peers up at the foliage overhead with the most pitiful interest. It is slowly weaned from a milk diet to the leaves of the cecropia, which the mother at first chews up for her offspring.

I once watched a young sloth about a month old and saw it leave its mother for the first time. As the old one moved slowly back and forth, pulling down cecropia leaves and feeding on them, the youngster took firm grip on a leaf stem, mumbling at it with no success whatever. When finally it stretched around and found no soft fur within reach it set up a wail which drew the attention of the mother at once. Still clinging to her perch, she reached out a forearm to an unbelievable distance and gently hooked the great claws about the huddled infant, which at once climbed down the long bridge and tumbled headlong into the hollow awaiting it.

When a very young sloth is gently disentangled from its mother and hooked on to a branch something of the greatest interest happens. Instead of walking forward, one foot after the other, and upside-down as all adult sloths do, it reaches up and tries to get first one arm then the other *over* the support, and to pull itself into an upright position. This would seem to be a reversion to a time—per-

THE JUNGLE SLUGGARD

haps millions of years ago—when the ancestors of sloths had not yet begun to hang inverted from the branches. After an interval of clumsy reaching and wriggling about, the baby by accident grasps its own body or limb, and, in this case, convinced that it is at last anchored safely again to its mother, it confidently lets go with all its other claws and tumbles ignominiously to the ground.

The moment a baby sloth dies and slips from its grip on the mother's fur, it ceases to exist for her. If it could call out she would reach down an arm and hook it toward her, but simply dropping silently means no more than if a disentangled bur had fallen from her coat. I have watched such a sloth carefully and have never seen any search of her own body or of the surrounding branches, or a moment's distraction from sleep or food. An imitation of the cry of the dead baby will attract her attention, but if not repeated she forgets it at once.

It is interesting to know of the lives of such beings as this—chronic pacifists, normal morons, the superlative of negative natures, yet holding their own amidst the struggle for existence. Nothing else desires to feed on such coarse fodder, no other creature disputes with it the domain of the under side of branches, hence there is no competition. From our human point of view sloths are degenerate; from another angle they are among the most exquisitely adapted of living beings. If we humans, together with our brains, fitted as well into the possibilities of our own lives we should be infinitely finer and happier,—and, besides, I should then be able to interpret more intelligently the life and the philosophy of sloths!

VI

MANGROVE MYSTERY

ONE day I found a hammock-form of roots, a maze of gentle curves which gave and braced, and, taking paper, looked to see if a mangrove had anything of interest to offer. At the end of three hours I slid painfully down into the rising tide, my unpenciled sheet fluttered off, and I went away with my mind in a whirl.

I rejoiced in Barnum's Circus long before I learned to write, but, if the first time thereafter, my mother had given me pencil and notebook with instructions to describe everything that took place in all three rings and on the stage, as well as the freaks, side shows and menagerie, my ideas would have been of equal clarity and inclusiveness as at my first mangrove séance. Above, around, beneath were interlacing trapezes, flying rings and rope ladders, liana nets and gaily painted poles, waving banners of emerald strung along the rafters, and high over all the canvas of the sky. And everywhere the performers—acrobats and leapers—worked mighty feats of balance and of strength; whiffs arose of strange and unknown creatures;

thrilling, tuning-up squeaks and umpahs came from hidden orchestras which surely soon must burst forth in full fanfare of breath-shortening music. Now and then a being would creep slowly past, (doubtless on his weary way to a long parade about some invisible arena), of such sight and form, that if raised to man's height would be a side show in himself.

But even at the first confusing survey, the mangrove stood out vivid and clear-cut. It had the aspect of a god, an Atlas, with feet firm planted upon earth, regardless whether currents of water or winds of air swirled about its knees, and with wide arms out—upward spread to the sky, upon which thousands of weaker beings found sanctuary. Some alighted for temporary rest of weary wings, others for longer periods, day boarders who came for meals or season residents who built their houses and reared their families upon the vibrant roots. And finally were those who knew no other world or scene, but, born or hatched upon the mangroves, clung to them until loosed by death. By their little body dropping to the water, they paid their final debt to Gravitation, returning to his implacable coffers this small meed of elevation-energy, which by grip of tendrils or of fingers they had possessed throughout their lives.

MANGROVE MYSTERY

These were all kindly, or at most indifferent folk, who if they gave nothing of value, did no harm. In a circus, the smiling faces of two acrobats who catch one another in midair may mask bitter hatred, a desire to swing short, or grip loosely; the story writers are fond of showing us the tragic sorrow obverse of the clown's grinning visage. In the sunshine and warmth of the mangrove tangle, behind the swaying leaves and bee-beckoning blossoms' fragrance are terrible strife and slow death. The splendid plant gives shelter and support upon its sturdy uplifted arms, not only to the fairy homes of humming-birds, but to parasites whose gratitude is never to cease strangling with inflexible coils, or, more insidiously, gently to insert living threads of death into the very heart of their supporter. Out of all this, how futile it seems to try to give any real idea of the marvel of mangrove life. At most we can hope only to arouse a worthy discontent, a disquieting desire also to go and see. For here are living tales, complete but as yet unworded, worthy to fill volumes of Carroll or Dunsany or Barrie or Blackwood; here are scenes needing only paper tracing to equal the best of Rackham or Sime, to touch the emotional gamut of Böcklin and Heath Robinson.

About ten thousand years before I filled this

fountain pen, some ancestor of yours and mine—our "touch of nature"—discovered that by building a house of piles out in a lake, he could thwart the wild animals which ever threatened him, and lessen the danger of a surprise attack from equally-to-be-dreaded envious or hasty-tempered neighbors. Few carnivores care to swim after their prey, and war canoes had hardly been invented. Such sanctuaries gave to families and to small tribes time to think, to invent new weapons, to seize new opportunities and to take better care of their babies.

Today, while pushing a canoe through the roots of the mangrove jungle, I thought enthusiastically of my pile-dwelling ancestors as I noted many exciting similes, and then paddled hastily back to the laboratory to see what botanists had thought about it. I found much of interest, but my mind was sobered, my imagination quieted. There was nothing of Swiss lake dwellings, but a very definite title of *Rhizophora Mangle*, and a casual remark of branches being supported by simple, vertical roots; it was put down that the petals were lacerate-woolly on the margin, exceeded by the calyx limb; but their delicate odor was passed by without comment; the living shifts of greens on the foliage, with the veins carrying shafts of parrot color over the back-

MANGROVE MYSTERY

ground of pale chrysolite—this was ignored; to the botanist the leaves were leathery, quite entire, obovate-lanceolate and blunt—a statement unquestionably to the point. Finally I learned that the astringent bark is employed for tanning, and I returned to my living mangroves, alias R. Mangle, wondering if too constant pondering upon astringent, unadulterated facts is not often efficacious in a sort of mental tanning. Our mangrove might yield a new harvest to us if we could choose a different contact of thought, clothing the fruit with the vital interest hidden in "one-seeded by abortion," and yet avoiding sentimental pleonasms.

However we decide to think of this plant, it is sure to be with admiration, for it stands out as a pioneer. Among earthly vegetation the mangrove is an aristocrat, a true dicotyledon, but it has dared to seek again the watery habitat of the lowlier growths, indeed of the very green algæ from which land plants originally developed. Like the penguin which has relinquished the aerial wing for an aquatic fin, or the seal which has encased its five fingers and five toes in flipper mittens, so the mangrove, while retaining all its badges of aristocracy, has returned to the haunts of the ancestors of all plants, from whence it can look calmly shoreward at the terrible struggle for life a few feet away,

JUNGLE DAYS

where every inch of soil is battled for, where the vigorous monopolize air and sunlight.

Such a radical change cannot be achieved without far-reaching adaptations and readjustments; the banker does not become a farmer merely by moving to the country, and every part of a mangrove shows delicate modes of meeting the strange new conditions as cunningly as the shift of muscles of a jiu-jitsu wrestler encountering an unknown opponent.

In the month of February, Kartabo mangroves are covered with flowers—and yet to a passing glance reveal no trace of inflorescence. Small and yellowish white, in irregular clusters of six to a dozen, they make no kind of visual showing, but their nectaries call to small trigonid bees in no uncertain way, and through the hours of sunlight the branches of the mangrove are busy marts of trade. Each cluster of blossoms becomes a corner grocery where the customers come for their buckets of nectar and packages of pollen and rush away without paying, or so they may think. But there are leaks in the pollen bags, and when they enter another blossom, the little stream of sifting yellow dust drifts across the entrance, a few grains or even a single one, falls upon the waiting pistil, and the bee has repaid for his bread and honey many fold and with compound interest. Its destiny fulfilled, the flower falls apart, the petal, lacerate-woolly margins and all, drifting off on the first tide. The ovary swells, two seeds form, and now comes the first adjustment, and we realize that in the botanist's dry remark "one-seeded by abortion" may be concealed tragic doom and a wealth of subtle meaning. No spear can be thrown straight which has twin heads and shafts, and so one seed shrivels and dies, and the other thrives and grows. What decides the fate of life or death we do not know. Some delicate balance, some subtle test of worth or lack takes place in every one of the thousands upon thousands of fertilized mangrove blossoms, and there is no appeal. The reason, as we shall see, is too vital, the target too difficult and treacherous for a thought to be given to unborn plants.

The problem of the next generation of mangroves is a serious one. The seeds are formed over an everchanging surface; soft, sticky mud giving place to strong currents, flowing first in one, then in the opposite direction; rough waves plough up the mud and splash against the stilt-like roots. No sticky secretion, no mere weight, no hooks or aerial wings will suffice for these seeds. From their natal branch high above the tidal area, some sure method of anchorage must be found to enable them to avoid being smothered in the mud, stranded on the shore, rolled into deep water or washed out to sea.

The method is the arrow or loaded dart, and the force is the energy of gravitation stored in particle after particle by the mother plant, as she drew up salts and water and elements, raising them sapfully from mud and tide, and condensing them into a solid, slender, pointed weapon capable of coping with all the difficulties of the new environment. But no seed alone can thus function, and in solving this problem the mangrove reveals itself as one of the most remarkable plants in the world. The lower forms of vegetation form their seeds and thrust them forth naked upon the world; the more advanced plants ensheath their offspring in swaddling clothes of protection against heat and cold, moisture and aridity. These are comparable to egg-laying creatures, with yolk and shell to shield the embryo from dangers. But the mangrove is truly viviparous, and the embryo seed remains attached for months, nourished by the sap of the parent branch. Out of the pear-shaped head a root-like structure grows downward, often to a length of twenty inches and a width of one. Like an airplane bomb, or the deadly throwing assagai of the Zulus, the mangrove seedling is thickest

MANGROVE MYSTERY

three-fourths of the way down, and then tapers rapidly. With a weight of as much as three ounces and driving force generated by a height of twenty feet, the umbilical cord of sap may safely dry, the connecting sheath shrivel, and one day there is a dull little spatter of mud, or a splash of water, and the unconscious work of the bees, the months of slow invigorating by the parent plant are fulfilled. The seed sticks upright in the mud, propelled through even two feet of water to its goal, and immediately rootlets sprout and consolidate the anchorage.

I once blazed two dozen seedlings which seemed ready to drop, and three of these were loosed at low water, so that they fell unhindered directly into the mud. The others I missed and I can only surmise whether this is the rule; whether some subtle influence of moon or tide is not sufficient to cause the final separation. Such a stimulus would be of great value to the young plant and is no more improbable than the marvellous effect of the moon's rays upon the palolo worms of the sea bottom.

Let us for awhile forget the mangrove circus medley,—crab clowns, strong men of the ants, hairy wild tarantulas, prestidigitator opossums producing ten infant opossums from a single fold of skin, white elephant membracid larvæ, living statue

JUNGLE DAYS

lianas, frog barkers and lightning change lizards. Let us think of birds, or of a single bird.

I have seen more than a hundred kinds of birds among the mangroves of Kartabo, but a mere enumeration of these would be of little value and of no interest. And instead of selecting the rarest, most bizarre of tropical forms, let us choose the commonest, the most blatant, apparently the most ordinary bird, with average habits and usual traits; which is another way of saying that we have observed it casually, watched it with unintelligent inattention, and wholly failed to interpret its activities in the terms of their desperate significance.

A kiskadee flew to a root before me and called loudly. For a moment it was only a kiskadee, and hardly registered color or sound, so common a feature of the day was it. It was threatened with the oblivion of the abundant, the neglect of the familiar. In New York City on a day of slush and humid chill, with rush and worry and congested life, to hear the loud, certain call *kis-ka-dee!* from a cage in the Zoological Park was to thrill in every fibre, and to remember peace, and calm thoughts and vast quiet spaces. As the steamer moved up to the Georgetown stelling, *kis-ka-dee!* from a corrugated iron roof signalized the approach of another season

MANGROVE MYSTERY

of wonderful jungle existence. But from that first moment on, the kiskadees were ungratefully allowed to sink into the subconscious, while jaded, conscious senses strained after new forms and novel sounds.

Today, however, looking up from my canoe among the mangroves, I saw the bird as first I saw it many years ago—it became more than one among hundreds, it assumed a miraculous rejuvenation.

Its very presence among the mangroves was significant. To the eyes of all immigrants through the ages the mangrove and the kiskadee must have come first—the tourist on the last ocean steamer, dark-haired men of quaint Spanish galleons, Carib Indians in their dugouts paddling from islands of the sea, and the man whose stone ax I found the other day, squatting on a couple of vine-tied logs, drifting from God knows where.

Here on the very apex, the outermost root, marking the junction of the Cuyuni and Mazaruni Rivers—here a kiskadee perched and here it had built its nest. It was exciting thus to be able to fix a locality with almost planetary, or at least continental accuracy. I have felt the same thing when circling in a plane over the very tip of Long Island, or standing on the spray-drenched, southernmost

JUNGLE DAYS

boulder of Ceylon, or squatting on a Buddhist cairn on the verge of Tibet. Now I knew that even a small map of South America would show this very spot of mangroves and the exact perch of my kiskadee,—and the bird grew in importance.

To Northern appraisement, our kingbird is nearest to this tropical tyrant, except that the latter is even more wonted to man's presence. The kiskadee has nothing of delicacy or dainty grace. It is beautiful in rufous wings and brilliant yellow under plumage, it is regal with a crown of black, white and orange. But in life and caste it is decidedly middle class. It is the harbinger of the dawn, but so is an alarm-clock, and in regularity and blatancy of announcement there is much in common between the two.

The husky call crashes upon the ear soon after the bird is sighted, and from early times has caught the attention and been translated into human speech. I know not what the stone-ax man dubbed it, he may only have grunted and hurled his weapon at it, hoping for a morsel of food. The Arrowaks and the few remaining Caribs know it as *Heet-gee-gee,* and the Spaniards, prompted perhaps by the Jesuit Fathers, interpreted it *Christus fui;* to Dutch ears it became characteristically tangled up with *g's* and *j's, Griet-je-bie,* the French more cleverly phrased it with the onomatopoetic *Qu-est-ce-qu'il-dit?* or *Qui? Oui, Louis!* while the negroes laugh it into *Kiss, Kiss, me deh'*.

I leaned back in the canoe and watched my kiskadee through a lattice of curved roots. Within five minutes it gave me a hint of the living chains of life with which the mangroves abound. The bird left its perch and with a wild outpouring of screams and shrill cries flew with unwonted directness, straight out and up over the river. Its mark was a caracara hawk—a menial, degenerate, vegetable-feeding *Accipiter,* who, when eggs or nestlings offer, loves to be tempted and loves to fall! Swiftly after the kiskadee swept the next link in the chain, two humming-birds whirring past, catching up at once and buzzing about the tyrant's head, well knowing that this sturdy eight inches of feathers, alias flycatcher, so quick to cry "wolf" at every passing hawk, was far from being wholly guiltless in the matter of certain nestlings.

But this is only an occasional failing and we pass to admiration of other, more worthy attributes. The kiskadee, like most strong characters has a number of doubles and imitators; one has drawn a grey veil over the yellow breast, another has a wider bill, two are almost replicas in miniature, but they are all conventional in haunt and food. They all

live in the compound of the bungalow and search the air diligently for winged insects as their names say they should. But kiskadee has overthrown the traditions of his family. A kindred spirit to the mangrove, his quick eye has caught the advantages of aquatic isolation and so we often find him nesting among the outer growths. And having accepted the sanctuary of this strange amphibious tree, he has altered his habits in other ways. A grey-throated kingbird or a lesser kiskadee will often choose a perch over the water from which it gracefully swoops for flying ants and termites. But watch the kiskadee!

As a returning crusader flaunts the infidel's scimiter, and keeps silence upon certain ways and means and happenings, so kiskadee returned to perch, wiping from its bill the sordid taint of tweaked hawk's feather, and ready to explain the lost feather from its own crown as worthy mark of battle. Its next movement was significant of much of earthly progress and evolution—indeed an accumulation of similar achievements would be quite enough to explain my sitting in a canoe, watching the kiskadee with high-power glasses, and endeavoring to philosophize upon what I saw, instead of still pushing my body into pseudopodia with my erstwhile amœbic confrères in the mud be-

MANGROVE MYSTERY

cleverly phrased it with the onomatopoetic *Qu-est-ce-qu'il-dit?* or *Qui? Oui, Louis!* while the negroes laugh it into *Kiss, Kiss, me deh'*.

I leaned back in the canoe and watched my kiskadee through a lattice of curved roots. Within five minutes it gave me a hint of the living chains of life with which the mangroves abound. The bird left its perch and with a wild outpouring of screams and shrill cries flew with unwonted directness, straight out and up over the river. Its mark was a caracara hawk—a menial, degenerate, vegetable-feeding *Accipiter,* who, when eggs or nestlings offer, loves to be tempted and loves to fall! Swiftly after the kiskadee swept the next link in the chain, two humming-birds whirring past, catching up at once and buzzing about the tyrant's head, well knowing that this sturdy eight inches of feathers, alias flycatcher, so quick to cry "wolf" at every passing hawk, was far from being wholly guiltless in the matter of certain nestlings.

But this is only an occasional failing and we pass to admiration of other, more worthy attributes. The kiskadee, like most strong characters has a number of doubles and imitators; one has drawn a grey veil over the yellow breast, another has a wider bill, two are almost replicas in miniature, but they are all conventional in haunt and food. They all

live in the compound of the bungalow and search the air diligently for winged insects as their names say they should. But kiskadee has overthrown the traditions of his family. A kindred spirit to the mangrove, his quick eye has caught the advantages of aquatic isolation and so we often find him nesting among the outer growths. And having accepted the sanctuary of this strange amphibious tree, he has altered his habits in other ways. A grey-throated kingbird or a lesser kiskadee will often choose a perch over the water from which it gracefully swoops for flying ants and termites. But watch the kiskadee!

As a returning crusader flaunts the infidel's scimiter, and keeps silence upon certain ways and means and happenings, so kiskadee returned to perch, wiping from its bill the sordid taint of tweaked hawk's feather, and ready to explain the lost feather from its own crown as worthy mark of battle. Its next movement was significant of much of earthly progress and evolution—indeed an accumulation of similar achievements would be quite enough to explain my sitting in a canoe, watching the kiskadee with high-power glasses, and endeavoring to philosophize upon what I saw, instead of still pushing my body into pseudopodia with my erstwhile amœbic confrères in the mud be-

cleverly phrased it with the onomatopoetic *Qu-est-ce-qu'il-dit?* or *Qui? Oui, Louis!* while the negroes laugh it into *Kiss, Kiss, me deh'*.

I leaned back in the canoe and watched my kiskadee through a lattice of curved roots. Within five minutes it gave me a hint of the living chains of life with which the mangroves abound. The bird left its perch and with a wild outpouring of screams and shrill cries flew with unwonted directness, straight out and up over the river. Its mark was a caracara hawk—a menial, degenerate, vegetable-feeding *Accipiter,* who, when eggs or nestlings offer, loves to be tempted and loves to fall! Swiftly after the kiskadee swept the next link in the chain, two humming-birds whirring past, catching up at once and buzzing about the tyrant's head, well knowing that this sturdy eight inches of feathers, alias flycatcher, so quick to cry "wolf" at every passing hawk, was far from being wholly guiltless in the matter of certain nestlings.

But this is only an occasional failing and we pass to admiration of other, more worthy attributes. The kiskadee, like most strong characters has a number of doubles and imitators; one has drawn a grey veil over the yellow breast, another has a wider bill, two are almost replicas in miniature, but they are all conventional in haunt and food. They all

live in the compound of the bungalow and search the air diligently for winged insects as their names say they should. But kiskadee has overthrown the traditions of his family. A kindred spirit to the mangrove, his quick eye has caught the advantages of aquatic isolation and so we often find him nesting among the outer growths. And having accepted the sanctuary of this strange amphibious tree, he has altered his habits in other ways. A grey-throated kingbird or a lesser kiskadee will often choose a perch over the water from which it gracefully swoops for flying ants and termites. But watch the kiskadee!

As a returning crusader flaunts the infidel's scimiter, and keeps silence upon certain ways and means and happenings, so kiskadee returned to perch, wiping from its bill the sordid taint of tweaked hawk's feather, and ready to explain the lost feather from its own crown as worthy mark of battle. Its next movement was significant of much of earthly progress and evolution—indeed an accumulation of similar achievements would be quite enough to explain my sitting in a canoe, watching the kiskadee with high-power glasses, and endeavoring to philosophize upon what I saw, instead of still pushing my body into pseudopodia with my erstwhile amœbic confrères in the mud be-

low. This thought came when the bird fell from its root, plopped into the water, and with effort, and a bit bedraggled as to plumage, rose with a small fish in its beak.

The eternal restlessness of two of our pet monkeys, "Sadie" and "Holy Ghost," suggested to one of us the excellent definition of a monkey:—"An animal which never wants to be where it is," and this applied to habits and traits emphasizes the importance of the kiskadee diving after a fish instead of merely swooping after a passing insect: the wide beak, the fringe of guiding bristles, soft plumage, the examples of its relatives and the instinctive dictates of hundreds of past generations, all point flycatcherward, yet it chooses otherwise and taps a more nourishing source of food supply closed to its superficial imitators, nearer to its new home, and less dependent on sun and season.

In this, as in all similar cases, the vital interest lies not in the fact of the actual change of habit, but how it came to arise. It were easy in the comfort of one's study with eyes fixed on pencil and paper to devise the method of origin, clothing it with facile words. There come to memory the shrill chatter, the swift short flights, the trim, stream-line forms of midget mangrove kingfishers, tiny Isaak Waltons whose plunge, strike and return embody the perfection of piscatory art. How easy for the intelligent eye of the kiskadee to observe the mode of life of these little neighbors of the roots, to essay, to practice and to succeed! Or if this strains our credulity, let us take another sheet of paper and again logically explain the origin of the habit: a pursued insect falls into the water, the kiskadee swoops at it at the same moment when a minnow arises; the fish is unintentionally seized instead of the flying ant, the foundation of cause and effect is laid; and so, "dearly beloved," that is the way the kiskadee learned to fish!

For my part, I have not the faintest idea of how it began, in fact the little I have been able to ascertain, tends more to complicate than to clarify the problem, but there is one very significant thing about this flycatcher fishing. The Kiskadee Tyrant (*Pitangus sulphuratus*) in some of its several forms ranges from Texas to the Argentine, and from Guiana to Peru.

Many years ago in western Mexico I observed the Northern form of Pitangus plunging for minnows in an arroyo pool, later, in the Orinoco delta and in Trinidad the subspecies *trinitatis* fished for me in both places; during five separate visits to Guiana I have seen many individual kiskadees catching fish in widely separated localities, and I

MANGROVE MYSTERY

have heard of a similar habit in birds of Brazil and Argentina.

Now while some unusually adaptable or quick-sighted bird may learn a new habit, or a new variation of an old habit, it is quite another thing to imagine a similar spreading of it wholesale among the individuals of the species ranging over mountains, plains and islands throughout a continent and a half. Such an achievement is as absurdly improbable as the theory of a kingfisher tutor. We do not know how it has come about, but when it is made clear I believe that many other equally mysterious phenomena will be understood; why so many groups of hoofed animals quite distantly related, all began in past time to develop horns more or less simultaneously; why in hundreds of tropical lakes which never know spring, untold hosts of ducks and geese are, as one bird, stirred by something beyond themselves—as inexplicable and invariable as the magnetic needle; why a flock of birds in flight has no individual will, but is swerved and turned, carried aloft or settled to rest by some inclusive spirit of flock or species. All this is not recognized by any taxonomist, it is not explained by psychologists, it is hardly ever thought of by naturalists, but some day it will demand of our philosophy an explanation. When that time

JUNGLE DAYS

comes, I will understand the fishing of my mangrove kiskadee as now I understand only how much I want to know.

A strange city or shore or jungle, a new friend, or house or garden should always first be seen at night; should be glanced at, not scrutinized, listened to, not examined, wondered at, not studied. The glamour rightly born of dusk will then forever mitigate defects apparent in the glare of day, ashcans, thorns, thick wrists, oilcloth tiles or blight. But no studied plan led my feet to the mangroves on a May midnight of the wettest moon at full. Raindrops from distant Venezuelan storms, and others which had spattered upon the mysterious heights of Roraima had filled the rivers up to their brims. And now the pull of the moon had slackened, and gently let the liquid mass sink down. There was not a ripple, only an occasional heave and settling, more effective, more potent of cosmic energy than any crashing waves or surging bore. And I did not wonder that ancient man failed to connect the tides and moon, for here high overhead hung the great satellite, while before me the gravity pull of yesternight's moon was just relaxing.

The light was somewhat grayed with clouds, but quite bright enough for type, if I had not forgotten that there was such a thing; the mangrove world

MANGROVE MYSTERY

was oxidized, the leaves lost all their semblance to foliage,—the branches merely dripped dark, oblong sheets of tissue. The slowly sinking mirror stretched the completed curves of roots,—slits widening to ellipses, ellipses to circles, until suddenly the earthly halves were shattered upon the dull glisten of exposing mud.

I was perched upon the buttress of a small mora which had ventured far out beyond its jungle brethren, or had been long since isolated by encroaching waters. Behind me was a black palm swamp and the narrow trail between. Optically both were invisible, aurally they were clearly outlined. From the swamp came the cheery little voices of the black and scarlet leaf walkers, the cubee frogs of the Indians, snapping out their brief but vital message, and from end to end the white-collared nighthawks patrolled the trail, with short, silent flights, thistle-down alightings, and never-ending queries of *Who-are-you?* as distinct as though worded by human lips. I remembered my Brazilian frog who pursued my researches with his eternal *Why?* I looked at the moon and the water and the mangroves, I thought of my imperfect self and I knew that never in this world would I form a satisfactory answer to either bird or batrachian.

Beyond the outermost roots came the low thrumming of a catfish singing in the shallows, forced perhaps by the lowering tide from some moonlit feeding ground hidden from my sight. It ceased abruptly and like an aerial antiphony came a deep rumbling throb from a root at my right,—the call of the greatest of all tree frogs, a well-named *Hyla maxima*. Here night after night I had heard him and had tried to approach. But always he detected my lightest step and became silent. His is the resonant bass violin in the orchestra of a jungle night. At this moment from two miles away, a chorus of these great frogs rang clear from a distant swamp. For about three-fourths of the time the calls were perfectly synchronized, coming in great successive waves; *wahrrook! wahrrook! wahrrook!* Wahruk, by the way, is their Akawai Indian name. Then some batrachian with a poor sense of rhythm got out of tempo, and this threw all the rest into confusion.

Now that I had remained quiet for many minutes, the fears of the giant tree frog were allayed and he called, almost within reach. I examined every branch near me and at last saw the outline of his great goggle eyes, standing high above his inconspicuous head. I even distinguished a huge webbed hand, looking like a bit of splayed out moss, resting flabbily against a bit of bark. In five minutes he rumbled forty-two times, grouping his emotional reiterations in series of eight, with long rests between. Steadily I watched him, until without warning, in the midst of a deep-throated *wahrrook!* he leaped into mid air. Only it was not my supposed frog with the outstretched hand which sprang, but a shapeless bit of dangling lichen a foot away, my image reverting into moss and bark; a lifetime of carefully trained eyesight availed nothing, even in this brilliant tropical moonlight, when pitted against the dissolving power of a giant treefrog. He splashed into the water, reaching another mangrove root in two kicks, and vanished again. This was not maxima's usual habit of a creeping walk from leaf to leaf, now and then leaping to a higher part of the foliage,—and I waited, and wondered.

In front of me were several twigfuls of leaves, and just below two curved roots, one complete from trunk to water, the second lacking a few inches of crossing the arc of the other. The air was motionless, the water like glass, when I distinctly saw three of the leaves move to and fro. Then two more farther on, followed by quiet, then all waved simultaneously, as with memory of the breeze of the past rising tide, or anticipations of the breaths which would usher in the coming dawn. No other leaf in sight even trembled,—only these rocked and swung. Another vegetable miracle followed,—the shortened root began to grow before my eyes! I had recently measured and marvelled at a bamboo shoot which pushed steadily upward almost ten inches a day, but here was a mere root which had added six inches to its length in half as many minutes! Finally my dull eyes cleared, and as the detective stories say, there was solved the mystery of the frog's leap, the shaking leaves and the sprouting root; a snake flowed slowly along through the leafy twigs, over the arched root to its tip, and then, with its suspended body, spanned the gap between it and the next root. Long before I had even seen the moving leaves, the frog had sensed the danger and fled.

As I watched the root apparently grow thicker, then diminish, and finally again become a shortened segment, my memory pared down the moon, cleared the sky of clouds, held fast to the mangroves, but raised the flat lines of bordering jungle into rounded hills. The palms and dark water and cool tropic air were the same, but instead of the roar of distant howlers there came to the ear the joyous whoops of gibbons,—the wa-was of the deep Bornean jungle.

All this leaped vividly to mind because it framed

MANGROVE MYSTERY

the last time I saw a snake among mangroves. That time the snake was smaller, but its effect was of infinitely greater moment. I was hunting Argus pheasants, but had unwillingly allowed my interest to be temporarily distracted by two great apes, orang-utans, which I saw now and then, and which were remarkably tame. One of these, a small animal about half grown, invariably retreated toward the river bank, and then vanished. No matter how carefully I trailed the strange little being, every trace of him disappeared when I reached the mangrove fringe. One moonlight night I sat upon a mangrove root, compass in hand, trying to locate a distant calling Argus pheasant, as the correct lining-up of the bird would be sure to bisect its dancing ground. After I had sat quietly for a long time, something drew my eyes upward and there, high overhead, peering down at me, was the orang, chin on hand, leaning on the edge of his nest of branches. There was no fear in his glance,—he looked like a meditative, aged man, who would have been more in place leaning on a cane in a chimney corner, than on a frail platform of broken boughs in a mangrove tree. I gradually focussed my electric flash on his face and he blinked at the strange light. He mumbled with his lips as if talking to himself, saying strange tree-top things about huge fireflies which burned too brightly. Once he swept a huge hand across his face, then sucked a great, crooked forefinger and without moving his head, rolled his eyes upward at a passing bat.

I shut off my light and we gazed at one another in the moonlight, with interest, but without malice or suspicion, until suddenly his twitching lips drew together, and I saw his whole body rise and stiffen. I followed his glance as best I could, somewhere beyond me, and before long I saw a small snake climbing out of the water up one of the roots. I knew it for a harmless species and after watching it draw out its whole length of three feet, I looked upward again. Not a sound, neither snap of twigs nor rustle of leaves had come to me, but the monkey's nest was empty. I could see the branches more or less clearly on all sides for thirty feet, yet there was no hint of the great ape. The harmless little snake had sent him off in violent but silent haste into the jungle, whereas my presence had given him no apparent disquietude. He was absent the following night, but the second night was back and actually snoring before I came close enough to disturb him. I never saw him again.

VII

THE LIFE OF DEATH

WE humans stand upright, but we look straight ahead. So for a long time I was blind to the mighty expanse of branch and foliage of my giant tree. I had passed it often and now and then reached out and touched it, for its mighty girth fascinated me. My Indian hunter gave me its name, Etaballi, and my botany added the less harmonious *Vochisia*. But it was my ear which first led my eye upward to a deep resonant humming which filled the dim air of the jungle. The sun was clouded as I looked, but the air was aglow with a solid dome of color, a gigantic mound of clear gold which eclipsed all the foliage and made the tree glorious. Humming-birds and bees, butterflies and nectar-loving wasps were there, and their wings of feathers, scales or mica tissue churned the air, each with an individual note, the sum of which was a composite tone of wonderful quality.

Lizards and woodhewers scampered easily up the trunk, birds and insects flew where they willed,

JUNGLE DAYS

but I was bound to earth and by stretching could reach at the utmost only eight feet from the ground. I could kill any bird in the top of the tree, I could call myself one of the Lords of Creation, but that helped not at all in my wish to study this majestic jungle growth.

Day after day I watched new masses of flowers come into bloom. Finally, so hopeless seemed the outlook and so marvellous appeared the teeming life of the tree-top, that I directed two amiable murderers, who were trail-cutting for me, to fell the jungle Etaballi. It was late when they began and the wood proved as hard and tough as metal, so when the warder came for them they had made but slight impression on the giant bole.

Then using a brain far better for mechanical achievement than my own, we evolved a plan for surmounting these ninety feet to the first limb. The plan did what I always like plans to do—it combined the primitive and the sophisticated. With a bow and special arrow of an Akawai Indian, a slender cord would be shot over the branch, then a rope pulled over, and with boatswain's seat and pulleys the rest would be easy.

The following day was one of great import both to the tree and myself. Much has been written of portents and warnings, and if I should narrate all

THE LIFE OF DEATH

the inexplicable things which have happened to me near the street called Prophecy, no one would believe the more ordinary events which occur as I traverse the avenue of Science. But in this case there was nothing. I left my friend in the late afternoon, standing in majestic quiet, leaves hanging motionless, although, a few hundred feet upward, white cloudlets were scudding before a mid-heaven trade breeze. I had seen this friendly tree lashed in tropic storms, I had watched it by day and night; parts of five years of our lives had been spent together, and I had seen but not observed its towering form as long ago as sixteen years when I passed up-river for the first time.

I had left Etaballi in the dusk, with its glory of gold pouring forth a stream of honeysuckle perfume and I looked forward to my new experiment in the morning, having to do with scaling its height. In the night arose one of the storms of the early rains. I heard the roar far down the Mazaruni and looked out of my tent to see first Pegasus and then the Pleiades erased, when there sounded the patter of the first few drops, followed by the steady, long, audible lines of downpour. Once and only once there came a deep distant *kr-ump!* such as used to roll over the wide sands and drown the surf on the coast of Belgium when the Germans were vainly strafing to the north. This single sound, as of a subdued exclamation of some great god looking down upon the jungle, was the only hint of anything unusual, and no one could call a far-distant thunder mumble a portent.

Nothing is more pitifully out of place than a fallen tree. It is like a foundered, deserted ship with decks awash, covered with a maze of broken masts, remnants of sails and tangled rigging. Thus I found my Etaballi, brought low, but worthy even in the manner of its fall. Human murderers had nicked it, but the final surrender was at the demand of one of the natural elements, whose brothers had brought the tree into being and nourished it into maturity,—a stroke of lightning,—sister of the sun, the rain and the winds.

Down it had come, straight to the north and cut for itself a mighty glade. All other trees in its path, all stumps and saplings, had gone down with it, and where for centuries had been dimness was now clear sunlight and a great expanse of open sky. The surrounding trees leaned far outward as if looking down with some strange arboreal sympathy for their fallen comrade.

I walked up and down the mighty bole, I swung myself up among the high branches, and even from those crippled, dying limbs I looked down upon

THE LIFE OF DEATH

earth from as great a height as the summit of an ordinary tree. I began to realize that in the death of my great friend I might achieve intimacy with many unknown things.

At present all was silent except for the rustle of shrivelled leaves and an occasional deep groan as some overstrained mass of fibres gave way. If birds had been perched or nesting among its branches last night, they had fled; insects had been shaken off, or were now making their way to other trees, as rats swarm along a ratline from a sinking vessel.

I left at once and did not return for two weeks. After that I spent an hour or two of many days with the fallen tree, and if I could have had my way every hour of daylight would have found me there. I wrote and collected until my fingers and body ached, and gathered a mass of astonishing facts which, when digested, will fill many papers with a multitude of very true, but to the layman, very tiresome, technical observations.

But always there kept breaking through the mist of bare happenings, of actual blatant phenomena, glimpses of the dramatic and the romantic side of this little cosmos. For the tree-made glade became an individual thing, a veritable worldlet, and just as we go into a room and to our delight find new pictures on the walls and new books on the table, so here in my gladeroom no two days were alike.

While sitting quietly in armchair, straight back or lounge—for I had all to order among the branches—I was forever having my attention distracted from the business at hand of bark and wood to visitors who came to peer or hammer, to play or to carry on their courtship almost within arm's reach. My angular figure and neutral garments were apparently an excellent camouflage among the maze of branches, and creatures came close which would have fled at first glance if I had been standing in mid-trail.

Every class of backboned animal except fish came to my fallen tree, and I have no doubt that to the leafy pools far down on the jungle floor, the land-travelling minnows had already made their way. Tree-frogs leaped past on damp, cloudy days and lizards of a half dozen species crept about, lapping up flies and other small fodder. A green tree snake came one day, but soon turned and went back to the protection of the surrounding foliage. An event was when a mighty boa constrictor, seventeen feet at the very least, weaved slowly in and out of the tangle. When he stopped he became but one more lichen-covered liana. In full sunlight he

THE LIFE OF DEATH

rested his great head flat upon a limb, and for many minutes no branch was more lifeless. Then I walked slowly toward him. When a few feet away he raised his head, looked at me, reached inquiringly forward with his pliant tongue, and slowly flowed away. We felt and showed mutual respect and each preferred to look, and then dignifiedly to turn aside, I the richer for the meeting, for I could add admiration and a thrill of real enthusiasm at the sight.

Monkeys came, a band of impudent Cebus, who dared descend to the branch tips, to shake them, and with many simian oaths to challenge me to come on. I took one step in their direction, and they fled chattering. Birds were almost always in sight—great yellow-headed vultures who swept down out of mid-heaven to see whether my prostrate body meant death. Doves boomed, toucans yelped, and after the first week a berry tree ripened its fruit, and no hour passed without flocks of parrots screeching full-lunged and sending down a rain of pits. Humming-birds fought overhead and fell, locked together bill and claw, at my feet; flycatchers found my glade a happy hunting ground.

One morning when I made up my mind to let no outside sight or sound through to my conscious concentration on the doings of the little people of bark and wood, I was suddenly startled into utter forgetfulness of my work. Here in the heart of the South American jungle there were reproduced for me the steep hills and valleys of northern Yunnan and Burma—the smells, the colors, the cold eddies of wind from the Tibetan snows—all were recrystallized in my mind by the notes of silver pheasants. From the underbrush behind my seat came the unmistakable low, liquid murmurs, breaking unexpectedly into the thrilling cackling. I dropped everything, and fifty feet away found a pair of distracted motmots who could not make their full-grown offspring behave, and were voicing their shattered nerves in this amazingly pheasantine outburst.

Herein lies the threefold charm of the labor of a scientist,—its unexpectedness, its mystery, and the eternal march of its phenomena, approaching, occurring, and passing into ever-vivid memory.

After the first week of observation my methods of close study had so sharpened my senses that the tree seemed to me to have passed into a resurrection of renewed vitality. Out of its death had come superabundant life. It recalled an observation by a stout fellow naturalist of mine, Samson by name, made many centuries ago. Some time after he had

THE LIFE OF DEATH

casually rent a lion in twain, he returned to look at the beast, and "behold there was a swarm of bees and honey in the carcass of the lion."

No part of it from underground roots to shrivelled topmost foliage was free from a flutter and bustle of vibrant beings. Thousands upon thousands of lives would cease and their races become extinct were it not for the occasional death of such a jungle giant as this.

An hour of undiluted, blazing sun drove me back to the splintered stump for shelter. I walked around and around it and then mounted it and fell to studying the cross-section smoothed by the skilful ax-blows of my friends the dusky criminals. I counted carefully, marking every century with a smudge of ink from my fountain pen, and when I had reached the very heart, I stood up and looked at the mighty Etaballi with renewed awe. I felt as if I had been unduly familiar with a stranger who was suddenly revealed as some very famous, very great historical character. For when this huge plant first broke from its seed and took root in this very spot where I stood, Genghis Khan became emperor of the Mongols. When its first leaves struggled for light and air the Crusades were at their height; on the opposite side of the world troubadours and minnesingers were making music, and

JUNGLE DAYS

Columbus and his voyages were still three centuries in the future.

For many minutes I remained quiet, held in wonder at the long centuries of human achievement. Then I returned to the watching of the life of to-day. I saw the excited creatures coming over the ground, along tangled branches or upon swift wings, and I saw that they were marvellously equipped, forearmed.

As I pondered on these mysteries and watched a sliver of a beetle crawling on the bark, human history blurred, faded and passed from mind. When Genghis Khan reigned, the beetle's ancestors were doing exactly what he is doing; double the years and Attila was making precedent for his successors—and identical beetle slivers crawled over dead bark. Ten times the years of this tree take us back beyond human history, add twenty or one hundred times its length of life, when our forebears were fighting to lift themselves above the other beasts, and in all probability not the slightest change could have been detected in the color, size, shape or habits of the flat predecessors of the tiny beetle under my lens.

When the bark begins to loosen a whole world comes by day and by night to creep beneath, and begin all the mysterious rites and achievements

THE LIFE OF DEATH

which fate allots to creatures of the under bark. All are positively thigmotactic which, as I once explained, is having the irresistible desire to touch or be touched by something, above, below, and—a thigmotac's greatest joy—on all sides at once. Twice I have experienced this and found it very terrible; the first time when I crept out of Cheops by the ancient, rubbish-obscured robbers' entrance, when sharp bits of alabaster so held me for a time that I could not move, and my imagination pictured the whole weight of the mighty pyramid pressing upon me. Another time was near the end of an obstacle race on a Toyo Kisen Keisha steamer, when each competitor, after fifteen minutes of constant, exhausting stunts on three decks, had to creep through a long, canvas ventilator laid flat on the deck. Half-way through, with the second man at my heels, I felt the canvas tube become narrower where an old tear had been sewn up, and my shoulders, even when pressed together, held the tube taut. Lungs full of coal dust, my blood beating in my ears like turbines,—no danger from savages or adventure with wild animals which I could recall, had ever given me a more ghastly minute.

I returned from my first day at the tree with a dozen beetles, and from a glance at them pinned in my collection, I can with certainty interpret their respective walks or creeps or crawls of life. A number are thin, but one is so amazingly flat that I am preserving it carefully among a few choice wonders of the insect world.

It is a small beetle, black and shiny as a new jet bead. It is oblong, and only by the most careful scrutiny can the faint details of the head, wings and body be detected. They seem no more than surface scratches and put to shame the most delicate watch or Japanese carving. I turn the beetle sideways and he becomes a mere black line, less in diameter than the slender pin which supports him. The under surface shows a more complex maze of lines, marking where jaws, antennæ, legs and feet are stowed away. He is a third of an inch long and a fiftieth thick. But above and below he wears his skeleton outside—a solid sheath of dense, hard chitin, and if we conservatively allot half of his thickness to this external armor, we have a space one hundredth of an inch into which is packed in perfect working order muscles for spinning his wings, walking, twiddling his antennæ and grinding his jaws; brain, nerves, eyes and other sense organs, mouth, stomach and intestine, and, if a lady beetle, ovaries whose scores of eggs are brought to maturity, with an intricate apparatus for depositing them. On another day I caught a wafer of an

THE LIFE OF DEATH

earwig whose bust measurement compared with its inch length, would, translated into human height, make a person just two inches in thickness. All the compactness of these shavings of vitality, these slivers of life, is in anticipation of the death of such a tree as this and the subsequent loosening of the bark.

Other beetles are antitheses of the first one, each a tiny cylinder with every surface rounded and every organ curved. The outer armor is a rich, glowing mahogany with a scattering of golden hairs and an absurd tail-piece, round, blunt and jagged. I did not realize the perfection of this arrangement, until, during the second week, I came upon a whole flock of these little chaps in their tunnels. After dark a flash-light showed only a tiny shaft driven into the heart of the wood, surrounded by cores of white, chewed-up wood pulp, but the moment the light struck down the hole, the faintest of shuffling could be heard by placing one's ear close, and like magic the hole vanished. The inmate had somehow detected the unwelcome light and had hastily backed up and plugged the entrance with himself. Now, looking at the pinned insect, the funny, round, jagged end-piece, so silly and meaningless in itself, resolved into a perfection of adaptation. No one could jump his claim!

JUNGLE DAYS

Beetles like these are stolid folk, wholly lacking a sense of humor, and they go through life, deliberately, directly, with never a side-wise glance or a light thought. In all this they have much in common with turtles.

Quick as the beetles were to take advantage of the new manna, others were before them, and I believe the very first comers were small, flat, wingless roaches, which scurried away as I lifted bits of bark. Roaches form the conservative wing of the insect world, and have many characteristics of certain persecuted human races. They are found everywhere, contented with a safe, middle course of life, seldom aspiring to size or bright colors, never attacking or even defending themselves, or putting on side in their life-histories. Once a cockroach always a cockroach is their motto. They have no responsibility of grub or pupal stage, and from the Palæozoic Age, unknown millions of years ago, to the present moment when one scuttled from the flood of light which I threw into his refuge, roaches have changed but little.

After the roaches or with them, for they resent no company provided they are allowed to creep and thigmotac in safety, came the wedges and gimlets of beetles, and in the next two weeks successions of stages of these hard-backs. First all but invisible

THE LIFE OF DEATH

eggs, then pale grubs squirming about in the fermenting wood, and finally a dynasty when the bark catacombs were filled with groups of stiff little mummies.

I excavated the débris in a deep hollow in the tree which once had been a hundred feet above the ground, and experienced something of the thrill of those who delve into ancient cities. At the top was a layer of twigs and leaves shaken up by the concussion of the fall. An inch or two below I found many berry pits and fruit seeds and when I scooped out several handfuls there came to light a dried and shriveled carcass, unmistakable in beak and foot—a nestling toucan which had never lived to fly and yelp and pluck bright berries in the sunlight of the tree-tops. Down I went again, into the very bottom of this nest midden, and there came upon rotten chips and soft, downy feathers. Among them were two, broken, stiff tail feathers which could have come only from one bird, the giant Guiana woodpecker, almost half a yard in length, with bill of ivory, and plumage of black, scarlet and white. No one could tell whether these birds nested in this stub within the decade, or when Galileo faced the Inquisition,—for the age of the supporting limbs made such latitude possible.

Still another discovery was left in my aboreal

JUNGLE DAYS

palimpsest. I was crumbling up the wood near the top of the hollow stub, where, long ago, it had been reduced by heat, water, fungi and insects to a rich, dark, pulpy mass. Suddenly, over a tiny chip, a weird little face peered at me, and a minute millipede, scurrying past, pushed over the wooden screen and exposed the quaintest being in the world. It was a doll or mummy—even the most technical scientist would admit the first, for he would call it a pupa, which was what little Roman children called their dolls. Being an average pupa it was motionless, and, propped up by accident against the dark, red background, it presented a multiple personality,—one thought of angel, curate, banker, clown, simultaneously. Around its head was an absurdly perfect replica of a halo, then came two mournfully sloped eyes, dark brown, sad, stolid; just midway down their diameter two translucent shields curved across, giving the little being the appearance of peering over horn-rimmed glasses; mouth parts were encased in crystalline coverings, a mouth which drooped at the corners— one felt that nought in past experience or future hope could ever twist that expression into a smile. Palpi were draped in each side like the side whiskers of a financier of the 'eighties. The two front legs, bent, with tips touching and elbows out, were

THE LIFE OF DEATH

laughably, like the comic paper idea of a country curate with finger tips spread and touching, gazing sadly over his glasses at some regretted irregularity of life. Then came the opal-sheathed wings, sweeping around in a beautiful curve across the whole of the underbody, as in old prints of guardian angels. Finally the tapering body-segments and their tip, fashioned in projecting styles. A hasty movement of mine sent down a shower of bits of wood, and buried the pupa. Carefully I uncovered him in his deep dark cavern and as I removed the last concealing chip, my little mummy gave me an unexpected surprise. From the hinder part of his body gleamed two dull lights, shining with a strong, steady glow, and illuminating the magenta walls of his sarcophagus. No wonder the appearance of these little chaps recalled most remarkable trilobite-like pupæ which I had found years ago in mid-Borneo, which proved to be firefly larvæ. I forgot all the comedy of halo, horn-glasses and finger tips, and with a little awe and much enthusiasm I watched the greenish-yellow shine. In the egg there is the first faint kindling—a dim, evanescent, rush-light glow; and here in the pupa, although it would have to wait perhaps many weeks before attaining adult beetlehood, its little lamps were trimmed and steadily alight, burning low it is true, and without the lighthouse rhythm of flash and blackness, flash and blackness. Already it was preparing for the all-important responsibility when upon the illumination would depend the chances of a mate and the future of its race.

The light of fireflies is one of the few things in this world which merit the term *perfect*. A gas flame is only three percent efficient, developing ninety-seven percent of useless, invisible heat or chemical rays; the blazing glare of the electric arc is only ten percent of what it ought to be, and most astonishing of all is the fact that sunshine gives off only thirty-five percent of visible light rays. Unlike Stevenson's "Lantern Bearers" the glow, deep-cloaked within the body of a firefly is wholly lacking in heat; it is one hundred percent pure flame.

I returned to the loosening bark and found that close upon the heels of the beetles came thrips, although these stout little fellows preferred the high, arched, dead branches to the main prostrate trunk. Few people have ever seen a thrips, but those who can find delightful the sound of the world itself have part compensation. When the time comes and one has seen and enjoyed a live thrips or a thousand thrips, then life will have acquired a new molecule of pleasure. If I say the word comes thrippingly to the tongue, it is only because I have just been consorting with a host of thrips, and their joy of life, their apparent love of play is infectious. Thrips are among the lesser folk of earth and if one attains the length of a third of an inch he is a Goliath of a thrips. But this, apparently like everything in nature, is comparative, for a thrips barely a fifth of an inch in length may harbor two hundred parasitic worms, who doubtless consider their host as gigantic. These tiny creatures are peculiar in many ways, as for example in their name which is both singular and plural. Also for unknown, but comparatively long periods of time, male thrips are wholly superfluous both for the continuance of the race, or companionship, or whatever other functions gentlemen thrips may be fitted to perform. In loyalty to my sex I pass this by, thoughtfully but without comment.

In the sizzling midday sun I first became aware that the era of thrips had arrived at my fallen tree. It seemed as if the samisen cicada players and myself were the only things awake in the world. The bark under my eyes suddenly assumed a salmon hue and my lens showed uncountable hosts of minute, scarlet thrips, all doing a frantic, zoroastrian dance. They were slender bits of life, with nondescript head and a tapering body looking like a string of scarlet buttons. They ran swiftly to and fro on their six legs, holding the body high aloft or thrashing it from side to side. Sometimes a half dozen thrashed together, in some diminutive wild rhythm, or two circled around each other, or antennæd some thripian scandal. Under the shoulder of one bit of bark dust three infant thrips practiced thrashing (a good tongue-twisting phrase!) until I tired of watching. All these were larvæ, or rather immature thrips, scarlet and wingless. Now every young insect with which I have ever been acquainted had thought and action only for food, but here was a whole generation of thrips—all under age—dancing and whirling about and waving their wild tails for hours during the hottest part of several days. I thought well of thrips for this unique casualness.

Every now and then an adult thrips appeared, somewhat larger, glossy black with scarlet seams and four marvellous wings. As wings they seemed hopelessly inadequate, but as ornaments they had much merit. If a crow were to shed all his wing feathers and was provided instead with four, small ostrich plumes, we would not expect him to fly. A mature thrips sports four delicate feathers with narrow shafts and wide, soft fringes down each side.

THE LIFE OF DEATH

I was once astonished to see a bony horse hitched to a decrepit car, slowly traversing a cross street in New York City, and learned that it was a mere gesture, a childish fulfilling of certain legal phrases in order to hold the franchise of the horse-car line. I recalled this when I saw an adult thrips coming through the air, slowly, uncertainly, with dangling body and pitiful feather wings barely sustaining the owner. This too was a gesture, a needless effort, for he landed heavily on the same branch, quite exhausted, a few feet away from the point of departure. On foot he could have made the distance quickly and with little exertion. Again I admired the thrips, for as in his youth he had played and danced as well as eaten, so now in adult phase he made the beau geste—the pitiful clinging to the franchise of his volant ancestors. His wings might be dwarfed by disuse, frayed by degeneration, but he could still cast with shrivelled muscles a shadow of past achievements.

The coming of the thrips was sudden, their ways were inexplicable, their going wholly mysterious. One day there were uncounted millions. Shortly afterward, needing a new more notes on their activities I went out and found every one gone,—not a single one remained. In their haunts were growths of evil-looking fungi, semi-liquid drops of

JUNGLE DAYS

scarlet trembling on yellow stalks, and around and among these sinister growths crept vast numbers of extremely small mites. These—plant and animal —were in turn evanescent and lasted but two days, but the going of the thrips will never be explained, —whether by migration, poison from the omnipotent fungus, or, as with so many other peoples of earth, through enervating lives of ease.

By sense of smell I could tell that radical chemical changes were going forward in the fallen tree. At first the glade was filled with the tang of aromatic wood, the clean, fresh odor of new split plant tissues; then the sap became heated and fermentation set in. The first stages were unpleasant, musty and acrid, but finally a malty whiff developed, which during my hours of research, awoke exhilarating pre-prohibition memories. If my coarse sense could detect these successive changes, what staggering olfactory blows must have been dealt to the delicate flies which came with the first hint of ruptured plant cells. Unlike the beetles they undertook their business in life with an apparent joyousness, and like the thrips they all had an inordinate love of the dance. It is a strange thing that at carrion and decaying wood we find so much graceful and intricate action, such varied courtship, so much effort only indirectly concerned with the

THE LIFE OF DEATH

odorous maelstrom which has summoned them all together. The visitors to beautiful and sweet-scented flowers and fruit, on the contrary, come and sip and leave, without delay or distraction.

I soon realized that I could spend all my time for at least a year on the study of the flies alone which came to the fallen tree. For ten mornings there came hundreds of small marble-wings, which wave their two, parti-colored banners alternately about. I looked closer and saw that they were clustered in groups of six to twelve, or more usually seven to thirteen. All the fortunate ones who had secured a mate were busy every moment protecting her from roaming males. The female fly had very short legs on which she walked briskly about, searching for suitable crevices to deposit her eggs. Her mate, on his elongated legs, stalked just above her, apparently anticipating every move. The pair would progress by quick, short spurts until a wing-waving stranger hove in sight. No introduction or preliminary challenge was necessary. The new-comer rushed up and tried to butt the husband out of the way. The rightful fly would haunch his thorax and brace his legs, for all the world like a football player meeting interference. Running swiftly around, the assailant would make another attempt on the opposite side. Meanwhile the female, apparently oblivious of all this strife on the second floor, went calmly on her way, making the engagement very confused and ineffective by thus constantly shifting the field of battle.

We should emphasize this admirable, domestic preoccupation to the full, for otherwise it pains me to record a lamentable lack of Lucystonism. The lady flies seemed indeed to care little what might be the outcome of the battles. When, now and then, her faithful guardian was overthrown and pushed into outer loneliness, the new protector was accepted without demur. In fact her bark-searching position allowed her glimpses of little more than the ankles of her Lord and Master, and it must indeed be difficult to be deeply moved emotionally by choice of ankles alone.

The battling of the mates was as it should be and has been since the beginning of time—brave gentlemen waging war over the weaker sex, but what shall we say of another group of seven where the seventh was an ignored wall flower! The poor little virgin did not accept her neglect in humble resignation, but proved herself a militant feminist, and made one attempt after another to drag her more fortunate sisters from the protection of their towering mates. She was always rebuffed and the last I saw of her, she was washing her face

THE LIFE OF DEATH

and hands, fly-fashion, after an ignominious tumble into a thimbleful of dirty water, which is fly-size for lake. How I longed to tell her of a scene being enacted only a few inches away, where I observed the meeting of two lonely bachelors. They began a most terrific head-pulling contest, until finally they separated unharmed and quite exhausted, and went peacefully off, perhaps realizing that after all in their case there was nothing in particular to fight about.

From a fly's eye height I looked down the prostrate trunk with twenty or thirty groups of tussling marble-wings in sight, their earnest but futile efforts to injure one another very comic to my eyes, but to them as serious as only fate can be serious.

Other flies had very different ensigns and dances. In one the wings were divided lengthwise, the front half being black, the rear transparent. These wandered singly over the bark and as they went, they swung first to one side, then to the other, at each swing opening out the wing on that side. The movement was exactly that of a skater taking long, oblique strokes, and swinging his arms far out to the side (a simile which could have no meaning for any native of this country). When two flies meet they do the outer edge around one another, closing in to battle if of the same sex, or to courtship if of the opposite. Others are perky peacock flies, with head and tail lifted in a position of eternal alertness, who slither along without perceptible individual leg motion, going sideways or backwards with equal ease. Their battle technique is like that of the bulldog, leaping from a distance, but the ferocity of their intent far exceeds their power of injury, and they bounce harmlessly off each other. They remind me of

"Empusa's crew, so naked-new they may not face the fire,
But weep that they bin too small to sin to the height
of their desire."

The creatures who come to gnaw and chew the dead wood are only one component of the complex maelstrom of life, siphoned hither by the smell of sap and decaying bark. One day an army of white fungus tents sprang up on a rotting branch, and a foot away even my poor human sense could detect a mildewy odor from them. Hundreds of insects scattered far and wide through the jungle, to whom the infinitely more powerful sap smell had meant nothing, were now vitalized into instant action, and there came into existence a whirlpool within the maelstrom. Great wine-colored beetles and smaller ones of various pigments, gathered in scores, dancing flies which were never seen on bark

THE LIFE OF DEATH

or carrion were summoned, and strange short-winged beings with scarlet tips to their slender bodies which they waved in mid-air like mock torches. As I knew from past experience the delicate, lace umbrellas would last only three days, and I watched with interest the race which these vital beings ran against time. No tunnels or mines for them, no prolonged courtship, but a quick mating and depositing of eggs which became grubs or maggots almost on the instant. Two days later, grubs were eating and molting with frenzied haste, and on the third day, when their nutritious shelters blackened and melted away, the larvæ dropped with them into the mat of leaf mold beneath.

The dilettante flies of the fungus puzzled me. Theirs were aerial dances, and for hour after hour they swung and feinted, swooped or hung like motionless motes. This mystery was solved when I took a number of the beetle pupæ to the laboratory and confined them in a glass observation dish. In a few days, instead of beetles, out came dancing flies. No wonder they had no need of haste; as parasites they could batten at leisure on others' labors. I looked askance at the rich regard of life and the new generation granted to what my Puritan forefathers would have decried as sinful, ungodly gaiety.

JUNGLE DAYS

Returning again to my bark I found a hundred similar cases. Spiders and wasps and many other enemies were gathering. Day by day the chains of life were forged longer and longer. Within my first week at the tree I could write the following from direct observation:

This is the bird
That caught the lizard
That ate the wasp
That stung the spider
That sucked the fly
That killed the grub
The son of the beetle
That gnawed the tree
That fell in the storm at Kartabo.

Or to be more technically explicit:

This is the Attila
That caught the Cnemidophorus
That ate the Pompilid
That stung the Ctenid
That sucked the Tachinid
That killed the immature Coleopteron
The son of the Elater
That gnawed the Vochisia
That fell in the meteorological disturbance at Kartabo.

And so the wonderful adventure went on. It had happened a thousand thousand times, and for uncounted miles in all directions were untold numbers of these trees whose lives would sooner or later terminate. My Etaballi, whose roots reached deep into the ground, and more than seven centuries into time, was dissolving. Bark and branch, sap and heartwood, by the alchemy of life were being rekneaded into a host of lesser beings—crawling, flying, dull and brilliant, hard and soft, clever and stupid, and as these poured forth from crevice or tunnel, cocoon or pupa, and their gauzy wings dried, their armor crystallized into malachite or emerald, there confronted them enemies in every guise and form. And presently the substance of the Etaballi, translated into the bodies of the borers, was resurrected into spider, lizard and bird.

Now and then I turn back to my journal for May the twelfth, and read the sentence: "The giant Etaballi fell last night." Science, Religion, Philosophy—how clear all these would be if we could solve this one mystery. I had hoped for some faint clew to the meaning of it all. I left my tree for the last time certain only of the profound inadequacy of my human mind.

OLD-TIME PEOPLE

Part I—Fact

A VOLCANO in eruption and a jungle monkey—nothing can ever quite prepare our minds for the first sight of these. Neither the crude wood-cut of Vesuvius in our old school geography, nor the latest colored moving picture of Kilauea, adumbrates the awe of the silent, ascending line of smoke, or the nocturnal glow of fires, old as earth itself is old. Your canoe slips through the reflection of everhanging jungle, and you suddenly spy a little face peering out from the fronds,—a face wistful, serious, grave as with the weight of planetary responsibilities; and so human that you feel that somewhere in its past it too could tell of an Eden tragedy. If not an apple, it must at least have nibbled a berry of some little vine of self-consciousness. How unlike the immobile features of the deer and rodents and jungle cats is this sober, anxious little ego! And how vividly our orchid climbing days return when we see a family of bandarlog swarming up a liana. These miniatures of ourselves seem to climb

OLD-TIME PEOPLE

as easily against gravitation as we loll down hill with it.

This Guiana jungle is a strange and wonderful place when we think of it from the view-point of its monkey tenants. Their floors are swaying vines and bending branches, their roofs green waving fans and banners. Their nearer neighbors are humming-birds and leaf-winged butterflies, gaudy toucans and screeching parrots. Far up through skylights they catch glimpses of vultures, soaring a mile above earth, and yet with eyes so keen that an accidental headlong fall to earth of any little monkey would bring a score of hungry ghouls. Through the skylight, too, hurtles swift death,—harpy eagles, whose grip is the end.

The jungle sends up enormous trees, one hundred, two hundred feet, among the branches of some of which fifteen hundred generations of monkeys have gambolled. If these stood like oaks in a meadow, isolated and alone, the four-handed ones would perish or have to take to the ground. But lignum vitæ rather than arbor vitæ should be the simian's password, for the vines which bind together the whole tropical forest are the way of life of the monkey. By means of the untold fathoms of ratlines and suspension bridges, tight ropes and ladders, these jungle people can range for thousands of miles without ever coming to earth, living in the realm of orchids and birds' nests, of sloths and tree lizards.

Their very name has come to be a byword, although, like their physical bodies in past ages, it is bound to us etymologically by monna and madonna. We laugh at their comic little faces and ways and, if we are incurably fanatic or quite egocentric, or fearful of what comes after death, we indignantly deny all past kinship of a common ancestor. On the other hand, if we love the truth and have a sense of humor, we recognize that these little jungle folk have missed being human by some very little accident, being, but for the grace of some side-tracking, ourselves. And while we swagger upright and think of our brains with complacency, are we sure that all the advantage is on our side?

As with us, the whole of the lives of these monkeys is one long struggle against gravitation. They are born and weaned, they play and fight, they eat and sleep, in midair far from the ground, and only when death comes, do the tiny fingers relax and headlong they slip through fronds and leaves to the earth itself. This same eternal pull of earth holds us completely in thrall at birth, then we roll over, struggle to hands and knees and creep reptile-like for a space. At last we rise upon unsteady soles

OLD-TIME PEOPLE

and from three to seventy we walk or run, swinging our arms to balance us, frequently tumbling to earth again, exhausted after a few hours and sinking upon chair or bed to gather strength against another day of upright struggle.

The joys of climbing, of balance, of swaying limbs, of headlong leaps from self-earned lofty vistas, pass with boyhood for most of us. They are renewed for me sometimes when I mount the ratlines of a ship plunging through heavy seas, or in the first rush of a nose dive from high in air.

We cheat the power of earth with elevators, though to do so we must call upon the lightning or waters for aid. Instead of holding to clean-barked boughs, swaying aloft in the sunlight, we creep beneath the ground and dangle unsteadily from dirty straps. In place of plucking our fruit fresh from its native stem and eating it amid the green glow of its own foliage, we barter for its shrivelled pulp sealed in cans of tin. We gape at and applaud those of our kind who dare, upon tight-rope or trapeze, feats which any self-respecting monkey would smack her child for thus bungling.

The Capuchin, the bourgeois organ-grinder's friend, in past years now and then climbed our gutter-pipe and at the reminding jerk on his cord, pitifully doffed his little cap and took our pennies.

Here in his home we tame him and bind him to us with affection, so that with full liberty he chooses his sleeping box on our laps. He is silent, and gentle and serious like the coolies who work on the coastal rice-plantations.

This is, of course, merely generalization, comparable to the immortal description, "The French are a gay and polite people, fond of dancing and light wines." Anyone who has been a friend to creatures,—dogs, birds, monkeys or any other of our quaint companions in this curious world,—knows that individuals vary in disposition and temperament only less than what we are pleased to call the highest order, Man.

Some Capuchins are silent; we have known some whose garrulity tried our patience and our hearing. There was once a man who took a cage to the African jungle and so far reversed the usual procedure as to enter it himself, while the gorillas congregated outside,—or so he hoped,—to gaze on the strange sight. His purpose was to study the language of gorillas. One suspects that the vocabulary thus acquired would be chiefly of a scurrilous nature, but who is so lacking in a sense of justice as to grudge the apes a chance to get even at last?

We have acquired some knowledge of monkey talk, especially from our Capuchin pets. It does

OLD-TIME PEOPLE

not seem an extensive tongue but the same sound can, as with us, be given many different meanings by inflection, pantomime, or even facial expression. When one of our small Cebus friends is confronted by some terrifying sight, such as a monstrous iguana, he springs away precipitately, wide-open mouth expelling on a sharp breath a guttural hissing grunt. Engaged with us in a game of tag around the laboratory, he sometimes finds himself cornered; then he emits the same sound, but no one could now take it for an expression of fear. It is much prolonged, without the abrupt tone of real terror, and his white teeth gleam in his open mouth in an unmistakable grin as he capitulates and flings himself confidently into our outstretched hands.

One wistful little chap who was once a member of the laboratory family would sustain his part in serious discussion for minutes at a time. To open the conversation, one had only to approach him closely, look him in the eye, and smack the lips gently and repeatedly. To this he never failed to respond in kind, but much more rapidly than human lips could move, wrinkling his brows mightily the while with the effort of concentration, and occasionally varying his remarks by an emphatic shake of the head and a curious throaty chuckle with a falling cadence, which sounded for all the world as though he demanded briefly, "Whatcher got?"

Monkeys have bad dreams, nightmares that perhaps are shared by us. Often in the evening I have been distracted from some microscopic business in hand by a clamor from the compound, and going out have seen a pitiful monkey face, with frightened drowsy eyes peering anxiously for insubstantial bugbears, and heard small whimpers of allayed distress as nervous little hands clung to my solid and reassuring fingers.

Most Capuchins have in their repertoires some almost bird-like tones of clear twitters and chirrups, and, when they are particularly anxious to be noticed, a sweet call, Coo-coo-coo, whose blandishment it is difficult to resist. This same phrase, loud and prolonged is the call of the clan when widely separated in the jungle. It carries over half a mile.

The Beesa monkey, like the native Indian, is a silent mystery. Neither likes close confinement, and no emotion is shown by their placid, inscrutable faces. The young do not understand the strange new beings who have come into their lives, and soon pine away; as long as they live they are extremely affectionate, but mentally dull and timid.

Beesas are strange-looking beasts. The fur is black, very long and coarse, the tail appearing as

OLD-TIME PEOPLE

large around as the whole body. The face is purplish-brown, surrounded in the adult, with a great ruff of yellowish-white. The young Beesa is more frowsy and less judicial in appearance. They roam through mid-jungle heights, a single great male leading his harem of five or six females, while as many half-grown youngsters trail behind. As they climb from tree to tree, sliding down vines or scaling steep aerial ladders, they utter a low, abrupt, penetrating grunt or cough sounding like a faint, dull blow of wood on wood, which ordinarily would never be noticed among the rustling of leaves and the occasional thump of a falling fruit or dead branch. When alarmed they slip away rapidly, and so short are their legs and so long their fur that they seem to flow instead of walk along the branches.

The squirrel monkeys or sackawinkis are, next to the marmosets, the smallest of the Guiana monkeys. Their noses appear to have been dipped into an ink bottle, and their brains into spirits of ammonia. They are living springs, never running down, but withal sober and silent in their contacts with life and ourselves.

There seems to be in some respects a relation between size and intelligence, not only as in elephants and shrews, but in monkeys. The marmosets,—tiny, furry, nervous little beings, are very stupid, food and safety occupying their almost every moment.

The monkey of monkeys of this jungle is the big red Howler. He lives in families, and when the great male raises his head and in the light of early dawn sends forth his mighty voice, its reverberations are distinctly audible three miles away. His tail is long and full-muscled, and the bare skin beneath its tip has lines and cushions which tell of things forever lost to us. The color of the long, silky hair is that of the gold nuggets in the streams which trickle through the jungle far below, and the emotions of our tame young Howler are those of a very young child,—he is curious, timid, resentful, excitable, greedy, affectionate, serious; as fond of lifting his voice in anger or joy as a negro at a revival and as volatile as a twenty-four-hour thermometer chart in a desert. Jungle monkeys, and an active volcano,—see them before you die, or you will have missed two splendid thrills in life.

Part II—Theory

A little monkey climbed down a swaying vine, hand over hand, until his face was close to a quiet pool of sweet water. The day before at evening, he had done the same thing. His mother and his ancestors for generations had done likewise. And

OLD-TIME PEOPLE

always they chattered at the monkey they saw in the water, and finally in anger snatched at him, and their little fingers troubled the water and the monkey vanished. Then they drank eagerly, turned quickly, and clambered swiftly up to rest.

Today the little monkey began to chatter, then stopped. He moved, and the monkey in the water moved. He brushed away some hairs from his face and the water monkey. Then something happened. He stopped chattering and peered again and again at the face in the water. He put his little paw over his eyes and slowly took it away. Then he forgot his thirst, raised his head and gazed fixedly before him, wrinkling his forehead and remaining very quiet. And the more distant his gaze, the less he seemed to observe, and the deeper became the wrinkles.

The night came quickly and the tragedies of the darkness began. The little monkey had long ago forgotten his momentary abstraction and was curled in a slumbering ball high among the dense foliage of a jungle tree. . . If there is such a thing as prophecy; if the first beginnings of great and momentous things make themselves felt abroad, then the cool night wind carried with it more than the scent of orchids and the calls of the night folk. It must have vibrated with the sense of the end of a great regime. The dominance of animals was tottering, the beginning of the end of earthly evolution. Something introspective had come to pass—a glimpse of the ego—a momentary flash of self consciousness. The little face in the water was not really another monkey. And the end of this realization was to be man.

But one such revelation was of no avail, and whether the little monkey was finally caught by his arch enemies—the serpents or leopards—or sometime slipped and fell into his pool we shall never know. But his memory can never die, for he was the first Seer; his eyes were the first to look Beyond and Within.

Then the new thing happened to great ape-like creatures. Day after day they would stop in their swift, hand-over-hand swinging through the tree-tops and gaze into space for a moment. These primitive *penseurs* were at a disadvantage, for when their less psychic brethren caught them off guard they promptly crept up and slew them. But relentless and remorseless as the waters of the open sea, these waves of abstraction rolled on. And like bits of drifting wreckage, came tossed and tumbled thoughts, dumb and inarticulate, groping and quite inadequate for any use.

The first periods of self-realization were like

trances or obsessions, wholly subconscious and involuntary. For that which we have not conceived, we cannot intentionally formulate. With feet and hands clasped about branches, the great ape beings swayed back and forth in the ecstasy of day dreams. Then from the inward view, the inner sight with unseeing eyes of what they could not name, they came gradually to look again upon the outer world. And now was wrought the great change, for linked ideas flashed upon their confused brain, twin stars of thought which in their grand-apesons might evolve into knowledge of cause and effect, and the greatest of all things thoughtful-correlation.

Against single thinkers, the thoughtless ones could easily prevail. And all the more easily because in the beginning it was as it shall be in the end—the law of compensation allots brawn to one, and mind to another, as dominant attributes. This abstraction was a thing apart, and unlike all other changes which had come in the past. When one stumbled upon a new way of opening cocoanuts, or experienced witless facility in walking upright for a few steps, one naturally kept the knowledge to oneself. Why should any new-found ability be shared! But these disturbing, inexplicable trances often led to a greater interest in one's neighbor or one's mate.

JUNGLE DAYS

Ah, one's mate! One had not thought of this before, except as a pleasing something to be kept near one. Blindly one had captured it somehow and one felt that one would tear that fellow ape apart with teeth and sheer muscle if he came nearer one's mate; and if . . . but here some buzzing fly was sure to distract, or a troublesome itching of one's back which required one's whole attention, and then, . . . well there was always something else, or food or sleep.

Not only to the great bull apes came these lightning glimpses of self, but to the females. But there was a difference. The correlation was direct. The momentary loss due to introspection was all but negatived by the instantaneous return to the objective: a return which was like the ascent of the diver with his pearl: a swift recovery of consciousness leavened with the unfathomable mystery of intuition. And through all the throes of thought conception, when bull apes travailed with wrinkled brows and aching heads for the sustained glimmer which ever faded and died out, their mates went about, ambling on crooked knuckles, and their little pig eyes shot swiftly their message to one another—they understood.

They understood and waited quietly. And for this waiting they shall have naught but praise, superlative praise. For it is not difficult to wait in ignorance. Thus the crystal waits for its perfect growth: the seed for the century-delayed warmth and water. But with understanding to have patience: to feel, however dumbly and blindly, the future of equality, of splendid unanimity of interest and respect, and to play one's hopeless, inarticulate part and wait—this is very wonderful.

And this was the part of the female apes, and the ape women. And the difference between these was too fine for any written words. But as nearly as may be it was the difference between waiting, and waiting with understanding. And there were ape women when as yet there were no ape men for them to mate with. They followed the law and accepted any bull ape who broke through their subconscious restraint—that restraint and appraisement which worked for evolution a hundred thousand years ago—and will tomorrow. So the bulls continued to come wooing like great brutal things of lust and brawn. And the ape women, with a last sidewise glance at their sisters, went with them.

And the bull apes, they too obeyed the law, and performed the three functions of their life—they sought their food, escaped their enemies, and enjoyed their mates. But they also did a fourth thing equally important in the long run, which was hard-

JUNGLE DAYS

ly classifiable, because it was instinctive and its selfishness obscured by heredity. They killed every weakling, or crippled bull or disabled female. One great brawny female had to use tooth and muscle to save her baby. Thus for once the law failed. And the failure of the law was due to intuition. And this was the second great result of the vision of the Seer.

The bulls had made but little use of their new-found self-realizations. But now the ape woman fought for her babe's life and won. Weak and small he certainly was, but he possessed wonderful quickness, and every pursuit and attempt on his life was unsuccessful. And he grew up and became a failure as an ape. For he tired of catching flies, and scratching and sunning and sleeping did not seem to fill up all the hours of daylight. He played with stones and gathered them in heaps, and then fled. For at this point all the bull apes in sight, having forgotten yesterday's identical experience, rushed up, expecting that such labor must mean new-found food. Then he found hollow trees and beat upon them for hours with palm or stick. But he sought no mate, which was perhaps fortunate, for he would doubtless have returned maimed, or else been slain outright by the outraged female.

Then one day came to pass the third wonderful

OLD-TIME PEOPLE

thing. A great woman, who had left her fang marks on every bull who had tried to woo her, came shuffling along and joined the weakling. He fled only a short distance and then returned fearlessly. For deceit and treachery were still to be evolved, and when the mighty ape woman showed favor to him he knew that it was truth. He accepted her, and continued to fear the world and to potter about with his stones, and bright-colored blossoms, and his banging of hollow trees. Then he commenced making club-like affairs, and sat outside the burrows of small animals and smashed them when they appeared. And one day he smashed the head of a female ape, who, following the fourth law had attempted to slay him, the unbearable weakling. Her mate was roused to such a pitch, that his self-consciousness dominated and he hunted his victim down. And this was the end of the weakling, who yet had carried out his destiny.

When the great ape woman bore a child, it fulfilled the promise of the little monkey's first ecstasy. The prophecy of the night wind had come to pass. Here was balance of brawn and mind. Against his twin thoughts, his correlation, his weapons, his resources, opponents melted away. And this first ape man found ape women ready: waiting and understanding.

THE BIRD OF THE WINE-COLORED EGG

IN this life of ours it is the striking and startling things which attract our attention and the inexplicable which focus and hold it. A tinamou fulfills all these requirements, but thrills only one person in a hundred thousand, because that is about the proportion of human beings which ever sees or hears or eats him. Nevertheless, tinamous range over forests and pampas of such extent that the whole United States could be laid down twice upon them without overlapping.

Quail, partridges and pheasants are birds of the north and temperate regions, and we are all familiar with the part they play in the life of mankind—æsthetic, recreational, and commercial. The stress of competition or some innate constitutional barrier hinders the dominance of these terrestrial birds in the jungles of the tropics. In the area of research at my British Guiana laboratory, only a single small partridge has found and retained a foothold, and this is a very uncommon bird. In its low call-note, its arched-over nest and its dead leaf plumage, it seems thoroughly affected by the

THE BIRD OF THE WINE-COLORED EGG

great, lonely dimness of its unusual haunts, and an observant traveller could remain for months ignorant of its very existence.

Another group of fowl-like birds has solved life in these great jungles by taking to the trees, even nesting high up among the branches. These guans and curassows have retained the whiteness of eggshell but have reduced the number of eggs in a single laying to two.

In the abhorrence of the well-known vacuum accredited to Nature, the absence of terrestrial gallinaceous birds is compensated by the presence of tinamous, bob-tailed, sturdy running chaps, who defy all the dangers of the tropics and carry on their lives in the face of innumerable foes. To those few fortunates like myself, who have had opportunity to admire, watch, study, listen to, shoot and eat these birds, the substitution is emminently satisfactory.

Five o'clock in the afternoon of a newcomer's first day in the jungle apprises him of the proximity of tinamous—although if unaided by Indian or ornithological lore, it may be months before he knows to what he is listening. From its sweetness, his guess will never be far from some song bird, perhaps of beautiful plumage, and from its ventriloquial character he will have no idea whether it comes from high overhead or from right or left on the ground.

Little by little, year after year, I have gleaned a habit here, a peculiarity there, until at last it is possible to piece them together into a mosaic of sorts, a shadowy palimpsest of life history which gives us more or less of an idea of the voice and fears, the food and courtship, and the strange domestic relationship of the sexes. The most familiar of the three species occurring in the quarter of a square mile of jungle at Kartabo is the variegated tinamou. My Akawai Indian hunters know him as orri-orri or maam, rolling the *r's* like any Spaniard, and when referring to him technically I call him *Crypturus variegatus variegatus* (Gmelin). This, for a wonder, is appropriate when translated, and the variegated hiddentail is an excellent and distinctive name.

My first problem was to discover whether the birds which I heard calling every evening were the same individuals or whether these tinamous wandered casually through the jungle except when actually nesting.

By means of slight peculiarities in the call-notes, I was able in two instances to locate with certainty the home range of the variegated tinamou. One bird, a female as it ultimately proved, was always

THE BIRD OF THE WINE-COLORED EGG

to be found in one of two small snarls of lianas and underbrush. Any time during the night the bird could be flushed from this spot. In the morning about 5:30 she began calling, timidly at first, then with more assurance. As it grew light she left her retreat and moved slowly west across one of our trails and then turned south to several trees with fallen fruit. Here the calling ceased for about half an hour and then recommenced as she retraced her steps, turned west again and went on until I lost her in the maze of thick jungle. Her last call was given about seven o'clock. During the period of a full month she followed this identical routine every one of the eighteen mornings on which I trailed her, with a single change to a new feeding ground when the supply from the first gave out. On five evenings I found her back in the brush pile, when she began a new period of calling, usually beginning about 5:15 and continuing intermittently until nearly seven o'clock.

Before the beginning of the regular silvery, staccato trill, a single high, sweet, long-drawn-out note is uttered, of about two seconds' duration, followed by an interval of three or four seconds, when the call proper is given. Rarely, when the bird becomes suddenly suspicious, the first note is given alone, but almost invariably it is the precursor of the call. When the birds rise they are always silent, unlike pheasants, no matter how terrified they may be. On moonlit nights I have heard their usual call at intervals throughout the night, on cloudy days it is sometimes uttered at noon, while during no month of the year is the variegated tinamou wholly silent. The call is, of course, always given from the ground, and probably nine-tenths of the utterances occur between 5:00 and 7:00 P.M. and 5:30 and 6:30 A.M.

The first note is usually on F natural, and is very sweet and penetrating, with considerable carrying power, being audible for long distances through the jungle. Several times I have heard these birds across the Cuyuni River, almost a mile away. It is a characteristic vocal utterance of solitary birds which inhabit deep woods, taking the place of motion, elaborate plumage, pattern and color of birds which have more of a chance to communicate by sight.

I have, as regards the enemies of the tinamou, three times found the feathers or other remains of this species in the jungle, once accompanied by the tracks of a margay cat or ocelot, and again by the pugs of some smaller carnivore; another record is of feathers of a tinamou in juvenile plumage in the stomach of a spectacled owl.

THE BIRD OF THE WINE-COLORED EGG

Variegated tinamous are naturally timid birds with a regular system of escape. When flushed in deep jungle they rise with a sudden rush of wings and scale off for twenty or thirty yards. They then come to earth and freeze for ten or fifteen minutes. If, as rarely happens, their landing place is accurately located, either by actually seeing the bird descend or the leaves moving, it is an easy matter to approach quite close and watch the bird for some time. It never moves while under surveillance but stands like a bit of mottled jungle débris with its eye full upon the disturber of its peace. Nine times out of ten, the individual flushed evades all scrutiny or search. Even more than in the great tinamou, the plumage of this species merges with the jungle floor. There is no doubt that the birds unconsciously trust to their protective coloring, both at first in permitting a close approach and in freezing after the escape dash. When one is crashing through dense undergrowth, the birds escape by creeping silently to one side, as I have now and then observed when crouching and watching the progress of one of my party near-by.

Once I saw a bird collide with a tree-trunk and fall stunned, although it ultimately recovered. But I believe that such accidents, due to imperfect steering ability, occur more frequently with the large tinamou than with either of the small ones.

These solitary birds seem to have no especial association with any other creatures of the jungle; more than once I have seen them stop feeding and look up in alarm at the warning rattle of an antbird which had discovered me, but this recognition of the quality of alarm in other birds' notes is common to most of the jungle fraternity.

Small berries or fruits form almost the whole vegetable diet, many cherry-like with round pits, wild plums with oblong stones, hard acorn-like seeds and occasionally fleshy fruits without pits or seeds. All the food is procured on the ground, and the birds in company with agoutis have favorite berry trees, under which, at the season of falling fruit, they may be found evening after evening.

They are as solitary in their roosting as in other ways; they roost on the ground, or, as in two cases at least, on fallen logs a few inches up. Usually the choice of place is deep within a tangle of lianas and vines, from which the bird could not possibly take immediate flight. I have kept close watch on a bird, which eventually proved to be a female, through a brief period of intensive vocal courtship, and neither then nor afterwards did the tinamou fail each night to roost by herself in her solitary tangle.

THE BIRD OF THE WINE-COLORED EGG

There are only three months during which I have no record of breeding and these would undoubtedly be filled up if I had more thorough knowledge of the field under observation. The calling of the females during every month would indicate that there is no absolute cessation of breeding, as there is in the case of the large *Tinamus*. The males of these tinamous take full charge of the single egg and the subsequent rearing of the chick, and I have found a male, attended by a three-quarters grown chick, incubating a newly laid egg.

I should not like to make any assertion as to a single male taking charge of more than three eggs in succession, but from two-month-period reawakenings of vocal calling in the vicinity of a single nesting area, and the number of young secured or reported from that place, I am quite sure that three eggs, one after another, were incubated. It is interesting to note that the same female, judging from the break in a preliminary note of its call, in the time under consideration, underwent at least three other periods of song development in an area somewhat to the northward, and although I could never locate a nest or a brooding male there, it is probable that she was courting if not actually laying eggs for another male bird.

In addition to this instance, at the end of March

I have secured a male variegated tinamou with one-third of the juvenile plumage still on the body, incubating an egg with a week-old embryo, and twice I have seen half-grown young birds in company with a single adult, presumably the male parent. My earlier experience with these birds indicated the remarkable proportion of sexes of eight males to one female. I now have a much larger series for comparison, and of forty birds secured within the area under observation, thirty-two are males and eight females, a very exact proportion of four to one. This is probably the correct percentage.

Almost all of the usual calling is done by the females, while the more excited vocal courtship is wholly feminine. Only once have I ever heard two birds directly answering each other, and on this same occasion I had my first glimpse of tinamou courtship. The male (presumably) was perched on a fallen log near my hiding place, while an approaching bird (later proven a female) came slowly, by short quick runs, from a bit of open jungle farther west. In the intervals between runs she gave utterance to a veritable ecstasy of calling —the usual dignified, deliberate scale being run and jumbled together in an excited, high-pitched flood of tone. The male answered from time to time

THE BIRD OF THE WINE-COLORED EGG

with the usual call, quite unexcitedly. With perhaps several months of brooding cares behind him, and more to come, we can hardly blame him for a restrained, philosophical exhibition of emotion. As the female approached, her runs became shorter and more irregular, her body plumage flattened, the head and neck were raised almost straight, and with rapid, mincing steps, her body vibrating with the effort of the continuous notes, she zigzagged toward the calm recipient of her attention. An abominable ant-bird discovered me at this moment, and rattled and screamed his loudest. Both tinamous seemed to perceive me at once, the male slipped off his log, and the female rose in a sharp, twisting spiral and I shot her as she turned, to make certain of the presumed fact that it was indeed the females which did the courting.

A few weeks later I was hidden between two fallen logs waiting for a quadrille bird to return to its nest, when a tinamou walked into view,—jigged, I might have said, for the bird was stiff-legged, and taking little mincing steps which shook her whole body and scuffed up the fallen leaves. It was exactly the tremulous heel-walk of an East Indian dancer when, with motionless body, he moves, or almost floats across the floor with short, rigid, almost imperceptible jerks. The tinamou revolved slowly, and when her tail came around into view I could hardly believe it was the usual dull-hued species. The tail, or rather, the ten, loose-vaned feathers which represent this almost obsolete organ, were upright, thereby pushing up all the elongated feathers of the lower back and rump. Closely applied behind were the under tail-coverts and even the feathers of the flanks, which now, flattened and with much of their surface exposed, proved to be really brilliant in color. With a shaft of sunlight striking them they fairly glowed; the tips of the tail feathers were buffy brown, then came a row of rich chestnut, then two rows of pale creamy buff with semi-circular narrow bands, then a beautiful patch of variegated feathers, white-tipped, with broad black and russet-red bars, and finally the softer, black-banded flank feathers. The wings drooped, the tips nearly touching the ground, the beak pointed upward, and the rich cinamon breast feathers were puffed out.

Three and a half turns did the courting bird make before she pirouetted behind the second log. What followed I did not see. I knew that the least movement on my part would send the bird headlong. My quadrille bird subsequently returned, I learned what I wished about her, and then, stiff from a prolonged squat, I arose pain-

THE BIRD OF THE WINE-COLORED EGG

fully. Like a shot, the two tinamous were up and bludgeoned off. Not a sound had they uttered, and after the faint scuffling of leaves which continued for a few moments after the birds disappeared, I had no knowledge that any tinamous remained in the vicinity.

The proportion of the sexes makes it almost certain that these birds are polyandrous, although judging by the slender spatial and temporal bond between them, promiscuous would probably be the more appropriate term. The lack of spurs and the insistence of vocality indicates that courtship and rivalry are carried on in ladylike fashion.

Of six nests found within the quarter mile of jungle under observation, three were in dry, moderately flat jungle, two in somewhat swampy places, and one on a trail half-way up the slope of a low hill. They are apparently chosen without any thought of escape, for in three instances when the bird got up, it either struck against intervening lianas, or had some difficulty in getting away clear. There is little doubt but that the site is chosen by the male; the hen tinamou sticks too closely to her calling place, her feeding and roosting areas to do more than court the male and lay her single egg. Once I was sure of a second site being near a former one. I took an egg in a damp low bit of jungle and a week later flushed the bird from a new, well-formed, but as yet eggless hollow eight feet distant from the first. He did not, however, return after this second alarm.

No attempt is made to form a nest. Attracted by some unknown choice, a spot is selected, and is made into a home literally by squatting. If leaves and twigs and other jungle litter are beneath the breast of the bird, they are pressed down and form the sole lining; if not, the mold alone receives the pressure and is gradually rounded into a shallow form.

A single egg is laid at one time and incubated. There is little variation in the color, the surface showing an exquisitely delicate tint which is but poorly expressed in our English term of light purple-vinaceous. There are sometimes zones of lighter tint about the larger or smaller end, due to some physiological cause in the lower portion of the oviduct. I consider the color of *Crypturus* eggs as distinctly protective, much more so than those of *Tinamus,* whose turquoise sheen is readily seen against the jungle débris. As such it is at least one ameliorative factor in the risk of the small number, and the danger of the continuously breeding male bird. The birds always sit close however, and only when almost stepped on do they boom

"One wistful little chap"

The Tinamou

From a painting by Helen Damrosch Tee Van

THE BIRD OF THE WINE-COLORED EGG

up and away. Many an egg would go undetected if, instead, the sitting tinamou would creep stealthily off at the first hint of danger. The gloss of the egg is not quite as high as in *Tinamus,* but it is still far ahead of any other bird's egg with which I am familiar,—one of the most beautiful shells in the world.

Out of the observation area I have known three eggs of the variegated tinamou to disappear suddenly long before incubation was completed, but only in one case do I know the cause, when a herd of peccaries trod heavily over the nest and all the neighborhood, a few fragments of yolk-stained shell showing how a single crunch had provided some wild pig with a delicious mouthful.

Incubation lasts about twenty-one days, and I have two notes, one of my own and the other by an assistant, of nests being deserted twelve hours and twenty-four hours after hatching. The parent therefore has at least the precocity of his offspring to lighten his labors. We have secured two young birds of about two and five weeks respectively, feeding by themselves at a distance from the parent, so the precocity extends to the independent juvenile life, thus allowing the male to take up, unhampered, a new round of domestic duties.

The position of the chick in the egg is very

obviously an adaptation to facilitate shell-breaking. The neck and head are folded close to the breast and abdomen, while the right leg is raised far forward and sideways until the beak rests directly on the under side of the flexed tarsus. Pressure is thus brought to bear on the shell not only by movements of the head but the slightest effort at extension of the foot and leg automatically forces the beak in general and the egg-tooth in particular against the inner wall of the eggshell.

On June 9, 1922, a single egg of the variegated tinamou was taken from a nest on the ground in the jungle. It was light purple-vinaceous with the usual highly polished sheen, and as well as I could determine through the dense pigmentation, the embryo was five or six days old. The egg was placed in the incubator in a temperature of 100 to 103 degrees and dampened and turned regularly.

Sixteen days later the egg was pipped at ten o'clock in the morning. Within two hours the chick was out, partially dried and creeping about all over the incubator shelf. The down dried well, but not on the back and head until I put in a circular band of flannel, into which the chick crept and by rubbing around as it would under its parent's plumage, the dorsal down dried fluffily. There is no doubt that the young bird would never dry well

THE BIRD OF THE WINE-COLORED EGG

without the constant friction of the old bird's feathers during the first twelve hours after hatching. This condition of the down is apparently a rather serious thing, for when the down dries flat and matted together, it causes such irritation that the little chick wastes much time and strength in trying to preen the bad places. Even a slight thing like this might very well be a matter of life or death, at a time when every moment of learning to correlate eye and beak is of the utmost importance.

I observed that the banging of the incubator door caused instant fear reaction—the chick squatting at once, but no other observations were made until the following day at ten in the morning when it was taken into the compound in a vivarium.

Placed on the ground the tinamou chick twice showed fear reactions, then pecked of its own accord. I worked with it off and on all day, and at last it took four small pieces of worms. On the whole it was far less apt in learning to calculate distances than *Tinamus major* of equal age. This was so marked that I believe it to be another example of very delicate balance between necessity and practice. In *Tinamus* there is a single adult to look after a brood of six to ten, while the solitary *Crypturus* chick has the whole attention of its parent, so there is far less need for extreme precocity in this case than in the former. With only a single chick to look after, greater care will be taken, and more time devoted to feeding and guiding the offspring. In *Tinamus* the young are compelled to forage more on their own, having the disadvantage of only a fraction of parental solicitude.

Another characteristic peculiar to this species in comparison with the larger tinamou is its relative silence. The other chicks, or even one by itself, were always cheeping and calling, whereas this one uttered only very low calls and at infrequent intervals. Even these are given only when the bird is quiet and undisturbed, and seem to be more in the nature of content calls then otherwise. It is readily seen that it is important for a covey of chicks to keep in touch with each other by frequent calls, whereas a single chick following its parent could with safety do so in comparative silence.

The *Crypturus* chick learned the use of its legs and by two o'clock could make its quick, short spurts without falling over at the end. It never walked slowly more than a step or two, but usually after several futile pecks at the bit of worm which I proffered, if it heard a sudden noise, it darted swiftly one or two feet away and squatted flat. I tested it with various sounds and found I could

THE BIRD OF THE WINE-COLORED EGG

cry out loudly or clap my hands together near it without effect, but the least deep or hollow sound such as striking the glass side of the empty vivarium, caused it to jump and flatten. Its pecking, as in *Tinamus*, was always forward and downward at the ground, and its constant fault was to strike beyond the object aimed at. The chick was uncomfortable on a white handkerchief and scuttled to bare ground as quickly as possible. It pecked at worms and spiders much more readily on the ground, even when they were of the same color as their surroundings, than when they were laid conspicuously on light bamboo leaves or when held in the forceps.

I tried calls and whistles with no apparent effect, until I imitated the note of *Crypturus* itself. Like a flash the chick turned in my direction, ran six feet toward me, and crouched beside my foot. I tried it again and again, then summoned the members of my staff to watch. The shrillest whistle brought no response, but the very first note on F natural above middle C, attracted and held the little bird's attention, and the following notes brought it headlong. After such a reaction it was much more alert and willing to attempt another bit of food, and not only this, but its sense of direction was almost perfect. When I held my face close to the ground and called, the chick ran, not only toward me, but stopped at my mouth, although I had finished calling before it reached me.

This instinctive and perfect reaction to the call of the species, together with its disregard of the call of *Tinamus* and other terrestrial jungle birds, was wholly unexpected. I have known chicks of other groups to crouch instinctively at the cry of a hawk, or the alarm note of their own or other birds, but to recognize among many other imitations, the exact summons call, was very interesting and threw a new light on the instinct reactions of this very generalized type of bird.

It did not enjoy being in the hot sun, but ran with quick darts toward the shade. Like the other tinamou chicks it never showed the slightest fear of our enormously tall figures stalking about. In fact, if anyone passed while I was attempting to induce it to eat, it invariably rushed off and followed, and had to be brought back and started over again in food interest. Unlike the large *Tinamus* chicks no shuffling of hands or feet in scratching motions and sounds had any effect.

Like so many of the small creatures I have watched in the laboratory compound, the chick persisted invariably in working toward the east or

THE BIRD OF THE WINE-COLORED EGG

northeast. Again and again I turned it about and always it changed direction and started back. I place no special significance at present upon this, but present it as an interesting fact as applying to mammals, birds, reptiles, amphibians and even to armored catfish. When, however, I gave the parent's call the chick never failed to turn and run toward me regardless of direction.

While it learned to peck and swallow bits of food and quartz with fair accuracy, I could not give it the constant attention and encouragement which it needed, and it died on the third day.

For many years the tinamou was a glorious anticipation—a hope engendered by the accounts of travelers in the tropical wilderness. It is now not only a memory but a stimulation, for when the city presses too closely, when four walls suffocate as well as enclose, when people oppress as well as associate, then I go to the bird house at the Zoological Park and at five o'clock there seldom fails me a sweet, clear staccato of silvery tones. Body and soul, I am back in the Guiana jungle, with the cool night settling down, a distant howler clearing his throat, and a bass chorus of giant tree frogs rumbling across the river. Then the tinamou calls again and the world is reorientated.